A Guide to

ROCK ART

Sites

SOUTHERN CALIFORNIA AND SOUTHERN NEVADA

DAVID S. WHITLEY

MOUNTAIN PRESS PUBLISHING COMPANY
Missoula, Montana

Library of Congress Cataloging-in-Publication Data

Whitley, David S.
    A guide to rock art sites : southern California and southern
Nevada / David S. Whitley.
       p.   cm.
    Includes bibliographical references and index.
    ISBN 0-87842-332-X (pbk. : alk. paper)
    1. Indians of North America—California—Antiquities.
2. Petroglyphs—California.   3. Rock paintings—California.
4. Indians of North America—Nevada—Antiquities.
5. Petroglyphs—Nevada.   6. Rock paintings—Nevada.
7. California—Antiquities.   8. Nevada—Antiquities.   I. Title.
E78.C15W53   1996
709'.01'130979—dc20   96-25752
                                                 CIP

Printed in Hong Kong by Mantec Production Company

Mountain Press Publishing Company
P.O. Box 2399 • Missoula, MT 59806
406-728-1900 • 1-800-234-5308

For my parents,
Ed and Bonnie Whitley

Contents

Contents

List of Illustrations

MAPS

FIGURES

PHOTOS

Preface

Of the many things our Native American forebears left to our modern culture, none is more evocative or inherently interesting than their rock art—the engravings and paintings found on cave and cliff walls, and the earthen figures marking river terraces. Ignored for many years by professional archaeologists, rock art nonetheless has interested if not intrigued the general public, who have now become frequent visitors to the sites.

Yet the professional archaeological neglect of this art form has had an unfortunate side effect: in the absence of convincing, readily available but scientifically and anthropologically informed interpretations of rock art, a number of inaccurate popular interpretations have appeared. These assert that the origins of the unquestionably Native American creations can be traced to such disparate groups as ancient Celtic and Libyan mariners as well as extraterrestrials (all lost, inexplicably, in the middle of the California desert). Alternatively, the misguided popular interpretations have attempted to reduce the complex religious symbolism of far-western Native American groups to a kind of naive sign language. Basic familiarity with Native American art of any kind is enough to convince most people that rock art is more than the scribblings of lost foreign travelers, and it defies being reduced to trail markers leading to water.

Luckily, professional archaeological interest in this field revived dramatically during the 1980s, rekindling an interest that had been frustrated by difficulties in determining the age of rock art and the lack of clear guidelines for deciphering its meaning. This renewed interest has resulted in substantial recent discoveries about rock art—information that gives us a sense of its age as well as confirming, in dramatic detail, some early interpretations of this art, thereby greatly expanding our understanding of it.

This guide summarizes the recent professional discoveries and conclusions about rock art in southern California and southern Nevada. It seeks to inform those who maintain an interest in rock art but lack the time, inclination, or access to the professional journals and monographs where these new findings have been outlined in scientific detail. The approach taken in much of the recent research—and thus the source of the conclusions outlined in this guide—is fundamentally ethnographic. It is based on a detailed examination and analysis of interviews and studies conducted by anthropologists early in this century of elderly Native Americans from California and Nevada, as well as some more recent ethnographic interviews with living Native Americans. In a real sense, this guide is an effort to give a Native American voice to the interpretation of rock art in far-western North America. Though this approach to interpreting rock art may seem obvious and logical, it has been ignored by archaeologists for many years.

The reasons for this surprising neglect are complex and are mired in the intellectual history of the discipline. They need not concern us in any detail here, except to grasp the concept that fundamentally they result from two causes. First, archaeologists are trained to study artifacts and material-cultural remains of prehistoric peoples; they generally lack the background and inclination to study and analyze documents obtained from living peoples. Perhaps more important, the body of information necessary to understand rock art is scattered and hidden in an array of documents; sifting, collating, and analyzing them literally takes years. Simply put, the ethnographic record had not been analyzed until recently because no one took the time to do the work required.

The initial stages of this analysis are now completed. While there is much yet to be learned about Native American rock art in California and Nevada, we now understand the broad outlines of the reasons for its creation and its meaning. This guide should provide you an opportunity to learn about both this art form and the Native American cultures that created it. Perhaps our thoughtful reflections on rock art and the religious beliefs and practices that underlie it will improve our understanding of related aspects of our own culture and society.

There is, however, one further motivation behind this guide: By exposing more people to and educating them about rock art sites, our ability to preserve these and other examples of Native American rock art will be enhanced. As my friend Russ Kaldenberg, State Archaeologist for the Bureau of Land Management, has cogently argued, this can happen in two ways. First, the simple presence of responsible and informed visitors, especially at remote sites, will serve as a deterrent to vandals who may intentionally or inadvertently harm the art. Second, unlike some aspects of the natural and cultural landscape, such as desert tortoises and wild burros, rock art still lacks a vocal constituency that urges land management agencies to place a high priority on the preservation, protection, and study of these sites.

Yet as is often the case, the squeaky wheel gets the grease. In our current era of diminishing governmental budgets, it is all the more important that people concerned with the preservation of rock art sites broadcast their feelings to the agencies charged with caring for the sites. I encourage you to help the small community of concerned archaeologists by becoming more than simply a visitor to these sites. These sites need all the help we can give them, and each of us who visits them can serve, in some capacity, as a steward of these fragile Native American resources.

In preparing this guide, I have consulted with numerous archaeologists and land managers, both to identify sites that are open and accessible to the general public as well as to clarify details of the art in realms outside of my specific areas of expertise. I am particularly indebted to Russ Kaldenberg, Dan McCarthy, Hector Franko, Ken Hedges, and Boma Johnson, each of whom has been very generous with their considerable knowledge of California rock art, archaeology, and/or Native American traditions. I also thank Sally Cunkelman, Kirk Halford, Carol Rector, Eric Ritter, Stan Rolfe, Richard Shepard, and Fredrik Whitley, all of whom have patiently and graciously answered my questions about specific sites and regions or who helped me locate sites that, like the foreign mariners and extraterrestrials, somehow had gotten lost over the years. Further, I thank Dan Greer at Mountain Press, who has been a very patient and helpful editor; Harry Casey, who graciously provided an air photo of the Blythe intaglios; and the Santa Barbara Museum of Natural History, for permission to reproduce a 1902 photo of Carrizo Painted Rock. And finally, I owe a special thanks to my wife, Tamara, and my daughter, Carmen, both of whom have been my willing and cheerful companions while visiting these and other rock art sites in many parts of the world.

UNDERSTANDING ROCK ART

A hiker walks down a dry desert canyon. Time and weather have discolored the canyon's basalt walls, turning them a deep, shiny blue-black. The hiker stops and unshoulders her backpack. As she leans into the shadows to drink from her canteen, she glances up at the dark cliffs. There on a flat section of canyon wall she sees a design engraved into the rock. It is a petroglyph. She ponders the image: Who made it? When? And, perhaps most important, why?

Many of us have had an experience similar to our apocryphal hiker, perhaps while simply enjoying the solitude of the outdoors or while actively looking for pictograph (rock painting) and petroglyph (rock engraving) sites. But whether serendipitous or intentional, the questions raised by such a discovery remain largely the same: What do the images mean? Is rock art an unknowable mystery, with its messages permanently lost through the passage of time, or can we unlock its secrets to learn more about the ancient Native Americans?

This guide answers some of the questions posed to us by the art itself and some you might then ask an archaeologist. Recent research has greatly clarified the purpose, meaning, and age of rock art, confirming a theory first suggested in the 1920s that its creation was intimately tied to vision quests conducted by shamans and ritual initiates. While many of the images in far-western North America are less than 1,000 years old, others may go back to the late Pleistocene—more than 11,000 years ago.

Keyed to thirty-eight sites open for public visitation in the southern part of California and southern Nevada, the goal of this guide is not only to show you the mystery and wonder of rock art in the Far West but also to provide you with an understanding of two aspects of Native American traditional culture—art and especially religion—often ignored or glossed over in other discussions about them. To understand the art, we must place it within its proper cultural context by considering the people who created it. We'll begin, accordingly, with a brief summary of the cultural patterns of Native Americans in the Far West, then we'll look at their religious practices and beliefs and examine the role rock art played in their religious rituals and symbolism. This background will help you to appreciate the messages conveyed by the art when visiting the sites.

FAR-WESTERN NATIVE AMERICAN CULTURES

With the exception of those groups living along the Colorado River, the native inhabitants of California and Nevada were hunter-gatherers who did not farm to any consequential degree. This limited the size and density of their population and required them to change residences seasonally. The native groups traced a path across the landscape as they followed ripening wild plant foods and animals according to the season. In many regions food became most abundant during the fall. The people collected and stored large caches of staples—usually acorns or pine nuts—at this time, allowing families to settle together in winter villages. As they used up the caches in early spring, the winter villages broke up and individual families scattered across the countryside to collect ripening seeds and tubers. The families remained on the move through the summer and recongregated in winter villages following their fall harvests. This yearly cycle repeated itself for many centuries.

Why didn't Native Americans of the Far West turn to agriculture, like the Pueblo groups in the Southwest? Native Californians and Nevadans did not practice rainfall agriculture because the primary native plant staples in the Americas—corn, beans, and squash—require summer growing seasons, and the annual precipitation in the Far West falls principally in the winter. In the Southwest, however, monsoonal rains coincide with the growing season, so agriculture there is more naturally successful.

The absence of farming in the Far West tended to inhibit the construction of the permanent year-round villages that typify native habitations in Arizona and New Mexico. The inhabitants of California and Nevada occupied a series of small temporary villages in various environmental zones at different times of the year within a group's territory. Since the villages themselves were impermanent, so too was the architecture; there was no advantage in constructing permanent dwellings when brush huts could be quickly assembled and just as quickly abandoned. The need for seasonal mobility, furthermore, limited the nature and quantity of far-western material culture—tools, implements, and artifacts—made and used by these Native Americans. Though we often equate wealth and success with the amount of *things* we own, for Native Americans living in the Far West the accumulation of too many things was counterproductive: when you must carry everything you own on your back, the wise person learns the virtues of economy and frugality.

Although the native peoples across southern California and Nevada shared many cultural similarities, some important differences also existed and are worth noting. For clarity and simplicity, let's divide the region into four cultural provinces, each of which is dominated by one or more different language groups. The speakers of any particular language typically divided themselves further into small governmental units within defined territories, but the speakers of a common language were no more the members of a common "tribe" than all speakers of English are citizens of only one nation.

Great Basin

The first cultural province is the Great Basin, which includes the Mojave Desert, the Owens Valley, and Nevada (see map 1). It was occupied by Numic speakers—a linguistic family comprised of the Northern Paiute (or Paviotso), the Shoshone, and the Southern Paiute (including the Chemehuevi). That Death Valley is in this region may remind you it is one of the driest in North America. Not surprisingly, Native American population size and density here were necessarily very low.

The traditional food staple in this region was the nut of the piñon, common on several types of pines. Piñon groves grow at the higher elevations in the numerous small mountain ranges of the Great

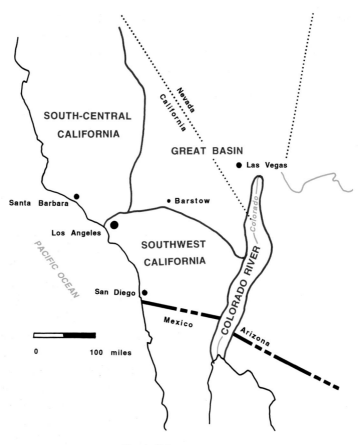

Map 1: Culture areas.

Basin. Family groups gathered around piñon groves when the nuts ripened in the fall. After gathering large quantities of the nuts, the families retreated to winter villages generally near permanent springs on the warmer valley floors. In addition to piñons, the families ate other seeds and tubers and supplemented their diets with the meat of rodents and rabbits, chuckawalla, and larger game, including mule deer, antelope, and bighorn sheep.

The inhabitants of the Great Basin generally organized themselves into village bands loosely governed by headmen who scheduled such food-gathering activities as the piñon harvest, communal rabbit drives, and antelope hunts. The headmen also settled disputes and, in times of war, organized raiding parties. But membership in any band was largely voluntary. A family stayed with a particular band as long as its leader brought them success and maintained harmony in the winter village.

South-Central California

The low native population density of the Great Basin contrasts sharply with the relatively crowded cultural province of south-central California, which includes the southern Sierra Nevada, the San Joaquin Valley, the south-central coast, and the Santa Barbara Channel region. This province contained the densest Native American population in the Far West, and perhaps the densest hunter-gatherer occupation ever to occur anywhere in the world. Present-day Santa Barbara, Ventura, and San Luis Obispo Counties were occupied by speakers of various Chumash languages. The San Joaquin Valley and Southern Sierra were the home of Yokuts speakers, while a series of smaller ethnolinguistic groups, including the Tubatulabal around Lake Isabella and the Kawaiisu and Kitanemuk in the Tehachapi Mountains, lived in between.

The dense population of south-central California reflects the particularly rich natural environments of the region. Acorns abounded in the inland areas, where this staple provided a reliable carbohydrate source. A series of sloughs, lakes, and streams in the San Joaquin Valley created a swampy zone rich in plants, fish, and wildlife. And the ocean currents and upwelling along the Santa Barbara Channel produced a bounty of marine life. These environments were so productive, in fact, that perennial villages of up to 1,000 inhabitants developed in some areas, rivaling the farming communities of the southwestern Pueblos in population density and permanence of settlement.

The high population density, permanent villages, and rich resource base contributed toward the development of chiefdoms in at least some portions of south-central California. A rudimentary class structure (such as chiefs, commoners, and outcasts) and an elite set of leaders who inherited political and religious powers typify a chiefdom. Villages within a chiefdom were organized into a political hierarchy, with a capital village, subcapitals, and so on. The primary Chumash chiefdom was ruled from a capital village near modern Point Mugu, in Ventura County, and was known as Lulapin.

Southwestern California

The southwestern California cultural province ranged from the Los Angeles Basin and the San Gabriel and San Bernardino Mountains southward across the Mexican border. Members of the Takic and Yuman linguistic groups—including the Gabrielino, Luiseño, Cupeño, Serrano, Tataviam, Cahuilla, and Tipai-Ipai (Kumeyaay)—lived there. This region represents a diverse range of natural environments extending from the coast, through the mountains, and on to the Colorado Desert in modern San Diego, Riverside, and Imperial counties. Accordingly, subsistence practices in this province varied more with location than by cultural or linguistic group. Beach resources such as shellfish were important on the coast, while acorns and piñons were exploited in the foothills and mountains. In the drier desert regions, mesquite pods and beans, piñons, yucca, and cactus fruits were important, along with a variety of seeds, tubers, and game.

Population density was somewhat lower in southwestern California than in the south-central region. The southwestern province had no large, permanent villages. Political organization within the various language groups of southwestern California was generally structured along the lines of clan-based "tribelets," which were autonomous political and ritual groups that owned a particular set of winter villages and typically occupied an established, definable territory. Members of a given clan tribelet believed they were descended from a specific common mythic ancestor. This genealogical descent, combined with the fact that most clan tribelets spoke slightly different dialectical variants of their language group, helped to distinguish one from the next.

Colorado River

The Colorado River cultural province includes the river valley and its surrounding terraces. Four different Yuman-speaking groups inhabited this region. The Mojave lived at the northern end, while the Quechan (or Yuma) lived along the southern end of the river, near the Mexican border, and the Halchidoma lived in between. The Cocopa lived on the river delta in what is now northern Mexico. Unlike other far-western groups, the Native Americans of the Colorado River were at least part-time agriculturalists, farming the rich river bottom to raise corn, beans, and pumpkins. They valued agriculture and the benefits derived from living immediately adjacent to their fields, so these Yuman speakers largely lacked the clustered villages of other far-western groups, as well as the need for extensive seasonal travel away from the river to other environmental zones. Their settlements consisted of extensive, sprawling, dispersed-family homesteads spread along the river, sometimes for miles.

Although this dispersed type of settlement might suggest a lack of social or political organization, in fact these Colorado River peoples were among the few far-western groups who maintained a national consciousness. In this sense, they probably constituted true tribes, loosely divided into a series of bands. They recognized a tribal chief and band subchiefs as well as war leaders. The war chiefs were particularly important, in light of the considerable raiding that was endemic to this region.

SHAMANISM AND VISION QUESTS

The relative frugality of far-western native cultures, when viewed from a Euro-American perspective that values material wealth over symbolism and aesthetics, led many early writers to pejoratively characterize these people as "primitive digger Indians." But when we consider the intellectual, artistic, and religious qualities of their cultures, we expose the ethnocentric bias of that label. Their successful occupation of such harsh environments as Death Valley obviously required a sophisticated knowledge of the region's geography and ecology. Moreover, any examination of Native American religion, ritual, mythology, and art reveals complex symbolic and metaphysical systems at work, even within those societies that lacked the outward trappings of material wealth.

The study of far-western North American rock art provides an excellent example of this by showing us how this art expresses these peoples' beliefs about metaphysics and religion in great detail. The cornerstone of religious beliefs in the Far West was shamanism, a form of worship based on direct, personal interaction between a shaman (or medicine man) and the supernatural (or sacred realm and its spirits). This interaction occurred when a shaman entered a trance. Often simply referred to as "dreaming," in a trance state a shaman could obtain supernatural power in the form of a spirit helper and receive the sacred talismans and songs the helper would impart. Typically, although not invariably, these spirit helpers were animal species with dangerous (and therefore powerful) attributes; grizzly bears and rattlesnakes, for example, were considered particularly potent helpers. Regardless of the species of animal helper, however, two points are critical: First, once a shaman received a helper he *became* that spirit, with his actions and those of his helper indistinguishable; second, animals that were important sources of food were rarely spirit helpers because it was taboo for a shaman to eat meat from the species of his helper.

The process of becoming a shaman was lengthy. Typically, a young person, usually a male, might receive a shamanistic calling by experiencing spontaneous trances, hallucinations, or vivid dreams, perhaps caused by illness. After many years of training, periods of fasting, and praying, the new shaman would retreat to a secluded spot for a vision quest, undertaken at night. If his preparation had been adequate and he was worthy, he would enter the supernatural through a trance and receive power in the form of a vision.

In Western neuropsychological terms, the shaman's trance was an altered state of consciousness in which he experienced aural, bodily, and visual hallucinations. A trance could be induced through a combination of isolation and sensory deprivation, physical stress (fasting, extreme exertion, or pain), drumming or dancing, and ingesting hallucinogens. Many people are surprised to learn that the primary hallucinogen used by shamans in the Far West was native tobacco, which is quite potent. They sometimes used other procedures and other plants—such as the dangerously hallucinogenic jimsonweed (*Datura wrightii*), for example, under tightly controlled circumstances—but rarely.

Depending upon which cultural province a shaman lived in, he might travel a great distance from his home village to seek a vision quest at a site used by many people, or he might choose a site near his village that he exclusively owned and used. In either case, vision quests were sought at "numinous" sites believed to be inhabited by supernatural spirits and, more generally, imbued with supernatural power. If vision quest sites were charted on a map, they would show how Native Americans perceived the distribution of supernatural power across the landscape. Native Americans believed power was spread over the countryside in netlike fashion. It was associated with caves, rocks, and permanent water sources like springs, lakes, and streams; generally, there is a correlation between such features and vision quest sites. Shamans sought places where they could enter the supernatural, and, metaphorically speaking, caves, rocks, and water sources were believed to be portals to the sacred realm.

A shaman might enter the sacred realm to use its power to cure, and sometimes to cause, illness; to find lost objects; to make rain; to "battle" evil shamans; and so on. Power could be used for good or evil; it was up to the individual shaman to decide whether it would be used for beneficial or harmful purposes. Nonshamans feared power because it was inherently dangerous, and they feared shamans because of the power they wielded.

In the Great Basin and along the Colorado River, shamans sometimes traveled great distances to particular spots where they could obtain particular spirit helpers or specialized supernatural powers associated with a given location. The best known of these sites is in the Coso Range north of Ridgecrest, California, in Inyo County — the primary locale for Numic speakers to obtain shamanistic rain power. Historical accounts indicate that shamans came to the Cosos from as far away as northern Utah. Another specialized vision quest locale exists at the extreme other end of Numic territory, in the Dinwoody region of the Bighorn Basin, Wyoming. Numic shamans traveled from all over the Great Basin to Dinwoody to obtain power in warfare. Of course, these same specialized kinds of power, and the helpers that imparted them, could also be obtained at other vision quest sites.

In contrast to the distances traveled by shamans from the Great Basin or Colorado River, in south-central and southwestern California the shaman's vision quest site was often near, or even in, his village. In these regions shamans conducted their vision quests in their home territory at sites referred to as "shamans' caches." Because the sacred realm was inherently dangerous to those who did not maintain the power to control it, nonshamans avoided shamans' caches, even if the sites were immediately adjacent to a village; people would go out of their way not to look at, talk about, or loiter around them.

Rock Art and Vision Quests

Immediately following a vision quest, the shaman would pray and concentrate on the visions he had received. When morning came, he would paint or engrave his visions on rocks at his vision quest site. Although there are some exceptions, as discussed below, much native Californian and Nevadan rock art depicts the *visions* of shamans while the rock art sites represent vision quest locales. The art created by a shaman was meant to preserve his visionary images for posterity; if a shaman forgot his visions, it was believed he would sicken or die. Shamans sometimes returned to the site of their first vision quest to revitalize the memory of that first trance, to renew their power, and perhaps to reenter the supernatural world for additional powers and spirit helpers.

We can develop fairly detailed interpretations of many rock art sites, as well as the art itself. We know the art reflects the visions shamans received at their vision quest sites because we have a series of verbal descriptions of shamans' vision quest experiences. Furthermore, we understand the human neuropsychological reaction to altered states consciousness; this is a key component to interpreting the shamans' trances.

Neuropsychology and Rock Art

The conclusion that much rock art in far-western North America depicts shamans' visions follows a similar discovery made by David Lewis-Williams and Thomas Dowson about rock paintings of the San (or Bushmen) in southern Africa. To understand the implications of trance-derived art, they conducted a detailed study of the neuropsychology of altered states. Fundamental to such a study is the fact that all modern human beings (*Homo Sapiens sapiens*, as opposed to *Australopithecus*) have equivalent neurological and physiological systems, regardless of whether they are African, European, or American in origin, and irrespective of whether they lived 10,000 years ago or today. Simply put, we are all "hard-wired" the same way, and therefore our neuropsychological systems react to trances in broadly similar ways.

According to neuropsychological research, people in trances typically experience four kinds of reactions. First, they have aural hallucinations; they hear things. The noises begin as whirring, buzzing, or tinkling sounds that, as the trance progresses, may be construed as voices, singing, or music. The second reaction involves somatic or bodily hallucinations, which might produce a feeling of weightlessness or flight or, alternatively, a feeling of heavy, leaden limbs. Some people feel as if their bodies, especially their heads, elongate upwards; the skin may feel like its growing hair or that insects are squirming across it. The third reaction is visual hallucinations—seeing things. Since sight is our primary sense, the visions look real, but they are in fact hallucinations that occur in the brain and optical system. The fourth reaction is termed a dissociative mental state: Although hallucinating, people in a trance usually retain sufficient mental rationality to realize they are hallucinating; they do not necessarily become psychotic and lose full sense of their rational or physical capabilities.

Although somatic hallucinations also have important implications for rock art imagery, visual hallucinations are of primary importance in understanding the art. Recognizing this fact, Lewis-Williams and Dowson developed a neuropsychological model of the mental imagery of trance states. It can help us understand how the graphics we see in rock art are intended to portray altered-states experiences.

The model has three components (see figure 1). First, it includes seven common *entoptic* ("within the eye") patterns. These are luminous geometric forms or light images that are spontaneously generated in the optical system. Staring briefly at a bright light and then quickly closing your eyes, or having migraine headaches, can generate these same entoptic patterns. They include dots and flecks, parallel lines, filigrees or meandering lines, grids and checkerboards, zigzags, nested curves, and spirals or concentric circles. Other entoptic patterns may also be "seen," but these seven are the most commonly described by people in different cultures who have described their trance experiences.

The second component of the neuropsychological model includes seven principles by which mental images are perceived during a trance. These include the simple replication of an image (entoptic or otherwise), the fragmentation of an image into its component parts, the integration of two or more images into a more complex pattern, the superpositioning of one image on top of another, the

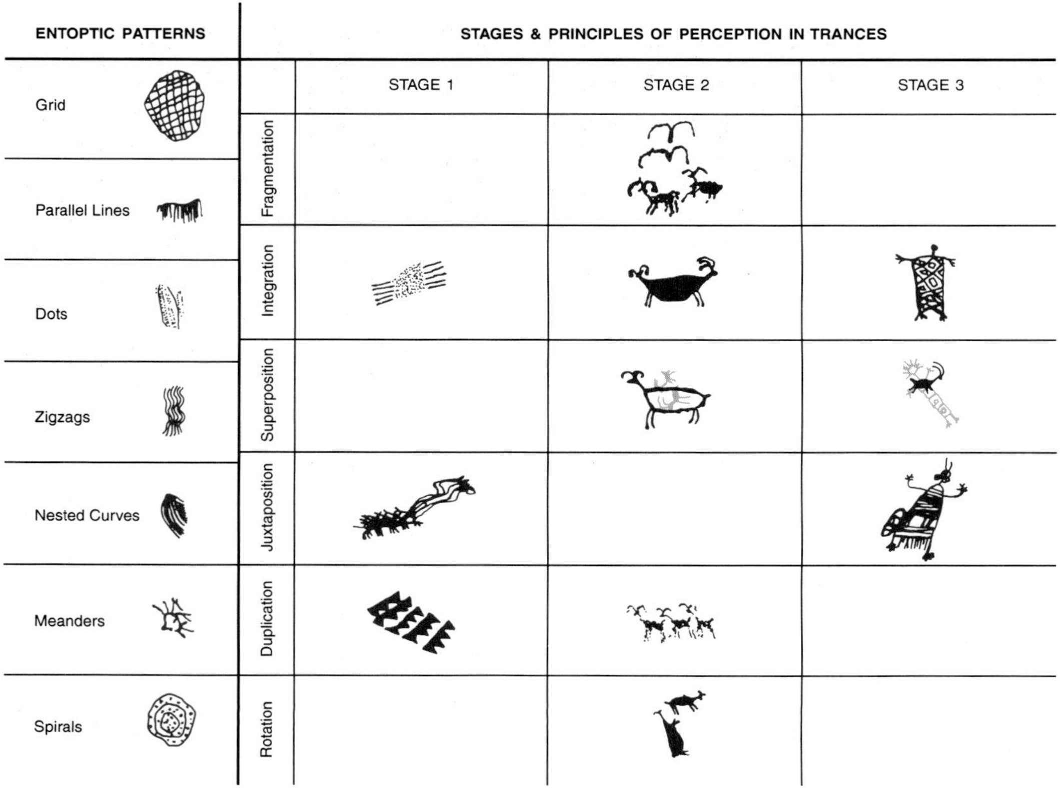

STAGES & PRINCIPLES OF PERCEPTION IN TRANCES

ENTOPTIC PATTERNS			STAGE 1	STAGE 2	STAGE 3
Grid		Fragmentation			
Parallel Lines		Integration			
Dots		Superposition			
Zigzags		Juxtaposition			
Nested Curves		Duplication			
Meanders		Rotation			
Spirals					

Figure 1: The neuropsychological model, developed by South African archaeologists David Lewis-Williams and Thomas Dowson, explains the effects of altered states of consciousness on rock art. The three components of this model are the seven common entoptic forms, the three stages of a trance, and the principles of perceptions, illustrated with petroglyphs from the Coso Range. Given that the art is intended to portray visionary experiences, this model explains many of the otherwise enigmatic characteristics of the art. (Scale: all different; in cases of superpositioning, red motifs overlie gray.)

juxtapositioning of two images against one another, the multiple repetition (or duplication) of images, and, finally, the rotation of images off the horizontal plane.

The third and final component of the neuropsychological model identifies three progressive stages of an altered state that influence a person's mental imagery. Initially, the person "sees" only entoptic patterns. Subsequently, these patterns may be construed as figurative images; the person interprets the geometric forms as familiar objects. Finally, the entoptics give way to representational or figurative images, such as a human, animal, or material-culture item — perhaps projected against an entoptic background or perceived alone. In this final stage of an altered state, the person in a trance may *participate* in the hallucination, perhaps even "transforming" into an animal, for example.

Cultural conditioning plays an important role in an altered state; a person's prior expectations can greatly influence the course of the hallucination. An example of such conditioning from Euro-American culture might be an expectation among alcoholics experiencing delirium tremens to see pink elephants; although it is somewhat of a joke, the association of pink elephants with alcoholism can condition the outcome of this special kind of hallucinatory experience. In the Native American cultures of California and Nevada, cultural conditioning greatly influenced the shamans' expectations toward their visionary experiences. Some of these expectations differed between the cultural provinces, as we might expect, and some were shared. The differences and similarities are discussed in reference to specific rock art sites subsequently.

There are several points worth bearing in mind about rock art and vision quests. The most important is that our understanding of the neuropsychology of trance clues us to the origin of many rock art motifs. We know, for instance, the art was made to depict visionary images, and we know that the types of mental imagery produced in an altered state is limited. But neuropsychology tells us nothing specific about what these images meant to their creators. Zigzags, for example, are common motifs in the art. Zigzags are also one of the most common of the entoptic patterns, and their presence in rock art is understandable in light of the neuropsychological model. Zigzags are exactly the kind of geometric pattern we would expect a shaman to perceive in his trance. To understand this pattern in a symbolic sense—and it did mean something very specific—cannot, however, be derived from neuropsychology alone. We must look more deeply into the ethnography of each of the cultural provinces.

Still, it is extremely useful to view far-western North American rock art in light of Lewis-Williams's and Dowson's neuropsychological model. It explains why geometric motifs—the entoptic patterns shamans "saw" during the initial stage of a trance—are so common in the art. It explains why some rock art motifs are fragmented, juxtaposed, superimposed, rotated off the horizontal plane, and/or integrated with other designs to create complex images: these are the principles by which mental imagery was perceived by shamans during their visionary experiences. It also explains why a single rock art site may display a combination of kinds of motifs, including geometrics (entoptics), representational forms on entoptic patterns, or simple figurative images, in all apparent respects fully naturalistic. These different kinds of motifs do not necessarily represent different styles of art, or the art of distinct cultures made at different times in the past, or necessarily even the work of different Native American artists. Instead, they may be the art of a single shaman, depicting the visions that he experienced during different stages of the same trance.

MEANINGS, MYTHS, AND METAPHORS

The meaning of rock art is, of course, its most interesting as well as its most elusive aspect. Although the meaning can be difficult to determine, it is neither totally idiosyncratic to each individual shaman-artist nor is it in all cases an unknowable mystery lost in the abyss of time. Instead, through a careful analysis of far-western Native American ethnography, we can apply some specific interpretations to many of the region's common rock art motifs.

The concept of *meaning* itself is subject to interpretation, so for clarification let's look at a cultural symbol we all know something about: the bald eagle. If several Americans were asked to interpret a drawing of a bald eagle, some would see it simply as the image of an eagle; others might say it represents the United States of America; still others might think of it as a symbol of strength, fearlessness, and majesty. None of these interpretations would be wrong, which underscores the first important point about symbols: they have multiple levels of meaning.

This does not mean, however, that symbols are too subjective to understand in any specific sense. If we asked our same group of Americans to consider the engraving of the bald eagle on the back of a quarter, most would think of it specifically as *the* bald eagle that represents the United States. This context of the eagle image—on a U.S. coin—gives it a particular symbolic meaning that is widely and readily recognized. Although the context of the art imparts a specific but shared meaning to the eagle image, at some point we also might wonder why an eagle symbolizes the United States. Strength, fearlessness, and majesty, as suggested above, are attributes of both this raptor and the United States; the eagle is a metaphor for the nation. We understand the logic and coherence of this symbolism: eagles *are* strong and majestic, and, presumably, so is the nation.

When we use animal symbols, whether for our nation, for sports teams, or for other things, we select species with attributes that we value and that we wish to equate with our organizations and products. We name sports teams the Falcons, the Rams, and the Bears; cars are the Jaguar and the Cougar; air weapons systems are the Sidewinder and the Cobra. In each case, it is apparent that there is a fundamentally rational symbolic system at work. Symbols are abstract concepts that play to our emotions, but we should not then conclude that symbolic systems are solely emotional and psychological. Instead, they may be fully as logical as mathematics, albeit slightly more abstract.

While our specific symbols differ from those used by Native Americans, the symbolic principles outlined above are universal and were employed by them as well as by us. Native American symbols, too, have multiple meanings, but specific meanings are revealed by the context within which they are (or were) used. Many Native American symbols, furthermore, are metaphors that link abstract concepts to visible characteristics or attributes of well-known animal species. Symbolic metaphors, thus, have been selected for use in specific situations exactly because there is an underlying rationale to their use. And because the use of rock art as symbols follows these same logical principles, we can begin to understand the meaning of the rock art of the Far West.

Myths and Art

Our fundamental clue to the meaning of rock art is the context of its placement on rock walls and panels. This context distinguishes rock art from images placed on baskets, painted on bodies, or drawn on sand, even though the same symbols may have been used in each case. Images painted or engraved on rocks had specific meanings not necessarily applicable in other contexts.

As noted above, many rock art sites were vision quest locales believed to concentrate supernatural powers, and they served as portals to the sacred realm. In this sense, the sites were far from neutral backdrops: They had a symbolic meaning in their own right, and this aids us greatly in interpreting the art. Furthermore, the ethnographic record is quite specific in identifying much of the art as shamans' visions of the supernatural. A critical factor in interpreting much far-western North American rock art is understanding the metaphysical nature of the supernatural realm that the shamans visited and depicted.

Metaphysical beliefs about the supernatural within native cultures of southern California and Nevada fall into two categories. For all but the Colorado River groups, far-western Native Americans believed the supernatural realm was a world geographically and temporally parallel to the contemporary, mundane world. There were divisions within the supernatural, in that most groups believed supernatural power came from the east, whereas the "land of the dead," another aspect of the supernatural, lay to the west. When a shaman entered the supernatural, accordingly, he traveled east, toward the source of power.

What is most important in terms of interpreting rock art, however, is that the supernatural realm was considered quite distinct from a third aspect of the sacred, which we can call "mythic time-space." This was the time and place in which mythic events occurred, before humans were created. In fact, the geographical locations of mythic events were well known and were associated with specific features on the mundane landscape. But the shaman did not enter mythic time at these locations during his vision quest, because to do so would have resulted in the witnessing or reexperiencing of mythic events. This was something shamans in south-central and southwestern California and the Great Basin did not do. In keeping with this fact, most rituals and ceremonies in the region involved demonstrations of the shamans' supernatural powers, while religious ceremonies, beliefs, and practices had little relationship to myths. Although myths were certainly acknowledged, repeated, and considered important, they did not figure in group ceremonies and rites.

Colorado River groups, in contrast, believed that the supernatural was precisely and solely mythic time-space. The primary purpose of their vision quests, accordingly, was the reexperiencing of the creation of the world during the mythic past, an event said to occur on a mountain peak known as *Avikwa'amé*, or "Spirit Mountain." It was from this first event and at this location that the Colorado River shamans received their supernatural power. The centrality of mythology among the Colorado River cultures is further demonstrated by its prominence in their rituals and religion. Typically, religious ceremonies among the Colorado River groups consisted of elaborate recitations of mythic song cycles, often requiring a whole night to complete.

These differing metaphysical beliefs about the supernatural are fundamental to understanding rock art, particularly because many people believe that rock art might be "illustrations" of myths, possibly depicting mythic characters. Along the Colorado River, this is apparently the case; the shamans entered mythic time-space during their vision quests and reexperienced the creation of the mythic world, so their art portrays mythic actors and events. But shamans in the remainder of southern California and the Great Basin did not enter mythic time-space; their art depicts different supernatural events and spirits. Remember, however, that in either case, shamans portrayed the supernatural or mythic events they experienced while in an altered state of consciousness. Even a graphic portrayal of a mythic being may have manifested itself as no more than an entoptic pattern, if this happened to be the manner in which this being was perceived by a particular shaman during his vision quest.

One final point about the mythic art of the Colorado River groups: The myths of these Yuman speakers were long, intricate, and complex. They did not translate easily to narrative illustration, such as might be provided by a picture book illustrating a biblical tale. It is unlikely that any of the Colorado River art depicts a myth in its entirety, or that the art was intended to be "read" as a narrative illustration of mythology. Instead, Colorado River rock art demonstrates a shaman's connection to a particular mythic being or event, and his art represents only the sacred patterns of the myths, not the individual incidents constituting the sacred tale.

Descriptions of Visions

Outside of the Colorado River groups, our understanding of rock art in the Far West can be guided by verbal descriptions of the shamans' visionary experiences and trance activities that have been recorded from Native American peoples. Although there is naturally some variation in their descriptions of the visions, typically the experience was said to begin when a crack in the rock at the vision quest site opened, revealing an entrance to a tunnel. (An important alternative beginning for a vision, discussed below, starts with a shaman entering a spring or pool and going underwater.) The shaman entered and walked down this tunnel, often encountering dangerous animal spirits along the way. Rattlesnakes and then grizzly bears are the most commonly described of the perilous animals the shaman crossed, but others such as mountain lions and spiders might also be included. These animal spirits guarded the supernatural, scaring away the unworthy and inflicting retribution on those who failed to obey ritual taboos. The concept of these guardians was most strongly developed in south-central California, where rock art sites often contain paintings of rattlesnakes and bears, indicating entrances to the supernatural.

Following the path of the tunnel, the shaman would eventually enter a large chamber and meet various supernatural beings, frequently the "master" spirits of animal species but also including ghosts (souls of the dead) and, in some cases, evil spirits. The shaman might also witness various events, such as curing ceremonies, dancing, warfare and fighting, or natural phenomena such as whirlwinds and lightning.

The beneficial beings a shaman met in his vision—and the songs, dances, and objects they gave him—were his source for, and the material manifestations of, his supernatural power; that is, they were his spirit helpers, his ceremonies, and his ritual talismans. An experienced shaman might also enter the supernatural to conduct supernatural activities related to his specific shamanistic powers, such as healing, out-of-body travel to distant locations, rainmaking, clairvoyance and finding lost objects, and, in some cases, sorcery, usually in the form of soul theft. A compilation of images and events described for the visions may include: human and animal spirits; evil spirits; feathers, crooked staffs, and weapons; fights; dances; curing rituals; footprints and tracks; insects; bright lights and stars; and, of course, the shaman himself in the supernatural.

Since we know shamans depicted the visions of their trances in rock art, it is not surprising that the images and events described above correlate closely with what is actually present in southern Californian and Great Basin rock art. This might suggest that we can easily interpret all motifs on the sites, but we cannot. Animal spirit helpers, for example, were only conceptually "animals." While they probably usually appeared in the shaman's trance in animal form, in some visions animal spirit helpers are described as male humans or as spirits that "shape-shifted" from human form to animal form during the trance. Depictions of what appear to be humans, in other words, conceivably may represent the shaman himself, a humanlike supernatural being, an "animal" spirit helper, or someone the shaman is attempting to heal or harm. Similarly, in some visions, powerful human spirits are described simply as "bright lights"—the entoptic patterns common to the shaman's altered consciousness. The complex nature of the mental imagery resulting from trances thus complicates our interpretive effort.

Still, and although this listing cannot be considered appropriate for all sites or cases, the following general interpretations are probably correct for many sites, remembering that the specific renderings of the motifs follow the principles of the neuropsychological model described previously.

Animal Figures: Descriptions of visions suggest images of mammals (including their tracks), reptiles, amphibians, birds, insects, and fish were often intended to portray animal spirit helpers. Certain animal helpers were associated with particular kinds of shamanistic power and specialized shamans. Throughout the Far West, for example, rattlesnakes were the helpers of Rattlesnake Shamans, who had special powers to cure snake bites as well as to handle the snakes. In the Great Basin, Rain Shamans were particularly important. Their specialized spirit helper was the bighorn sheep. Even the occasional horse-and-rider paintings or engravings may be understood in this manner: In the historical period, a special group of Horse Shamans developed to cure equine ailments. Their spirit helpers, naturally enough, were horses.

Dangerous species, particularly rattlesnake and grizzly, sometimes portrayed guardians of the supernatural realm rather than spirit helpers, so the relationship between animal motifs and helpers is not one-to-one. Lizards and frogs, as another example, crawl in and out of rocks through cracks or jump in and out of water, respectively, and this is analogous to a shaman's entry into the supernatural by metaphorically entering either a rock or a spring. These animals, therefore, were believed to be messengers between the mundane and supernatural worlds, and they were metaphors for the shaman's own ability to move back and forth between the mundane and the sacred realms. They might be portrayed on rocks to signal the locale as a portal to the supernatural. Insects, particularly centipedes, were strong spirit helpers, but such multilegged fantastical insects might also symbolize the dangerous powers and spirits seen in the supernatural, sometimes verbally described as "disease" or "swarms of insects." Similarly, headless or fantastical creatures might be the spirits, including ghosts, sometimes seen while in a trance.

Human Figures: With only an occasional exception, the human figures described in visions and found in rock art are male, reflecting the fact that shamanism was largely a male activity in this portion of the Far West. The figures may represent spirits in human form seen in the vision, someone the shaman intended to supernaturally aid or bewitch, or, probably in most cases, the shaman himself transformed into his supernatural aspect. In the latter case, the somatic or bodily hallucinations of a shaman's visions are sometimes portrayed by stretched bodies, upwardly elongated heads, or the partial transformation of a shaman from human to animal form. Similarly, because the shaman "received" his power songs, dances, and ceremonies during his visions, depictions of groups of humans dancing, or humans in a dancing posture, are common. The shaman's dance posture was arms lifted perpendicular to the body, with elbows bent at 90 degrees, exposing the shaman's "seat of power," his right wrist, to the sun, as is sometimes shown in the art.

Implements and Objects: Material-culture objects, including bags, weapons, staffs, and headdresses, are particularly common in Great Basin rock art and are sometimes present elsewhere in the Far West. These represent the power objects, costumes, and talismans shamans received in their trances; they were therefore the ritual objects the shamans used in their ceremonies and rites. For example, in the Coso Range area of the Great Basin, a number of the human figures are depicted with headdresses bearing the distinctly curved topknot feather of the quail. These headdresses were part of the special costume of the Rain Shaman. The prominence of weapons—like bows and arrows and the older spear-throwing board, or *atlatl*—furthermore reflect both the fact that these "weapons" were used in shamans' rituals and ceremonies and that the art and its symbolism are often oriented toward male themes and symbols.

Geometric Designs: Geometric motifs generally refer to the entoptic patterns experienced during a vision. As emphasized above, the simple fact that we can identify this type of trance imagery tells us nothing about its meaning. Furthermore, the fact that many of these designs are essentially idiosyncratic and unique suggests that their meanings may also have been idiosyncratic and therefore irretrievable without the comments of their shaman-artists, which are of course lost to time. Since we do not have records of such comments, and since the majority of the motifs at most sites are entoptic-geometric forms, the unfortunate conclusion is that we may never be able to interpret in any detailed sense most of the rock art in the Far West. This does not mean, however, that the meaning of all entoptic-geometric patterns are entirely lost.

Ethnographic data on three geometric designs, commonly repeated at rock art sites, helps us interpret their meaning. The first two entoptic-geometric patterns are the zigzag and diamond-chain motifs, both of which were universally identified in far-western rock art, basketry, and body painting as rattlesnake motifs. The logic of the connection between these motifs and the rattlesnake is straightforward: Zigzag is the track left by a sidewinder in the sand,

and the diamond chain is the scale pattern on the back of the diamondback rattlesnake. Zigzag and diamond-chain motifs, accordingly, can be interpreted as rattlesnake spirit helpers or, particularly in south-central California, alternatively as guardians of the supernatural.

The third interpretable entoptic-geometric pattern is the concentric circle or spiral, which is particularly common in the Great Basin. This was a visual representation of the whirlwind, a natural phenomenon shamans believed could concentrate supernatural power. The Great Basin Circle Dance mimicked the counterclockwise direction of the whirlwind because it too was believed to concentrate the power of the group when it was performed. The whirlwind was also thought to contain ghosts and, sometimes, to carry a shaman into the supernatural. Indeed, in south-central California the words for whirlwind and the shaman's magical flying power were one and the same: a whirlwind was a shaman using his power to fly. In the Great Basin, in a similar way, human figures are often rendered with concentric circles or spiral heads, symbolizing the shaman's ability to concentrate supernatural power and fly into the sacred realm.

Metaphors of Trance

One final area to discuss about the logic underlying the use of symbols in far-western North American rock art concerns two important points alluded to earlier: first, a shaman's vision quest experience during a trance was necessarily "otherworldly" and therefore inherently difficult to express verbally or graphically; and second, abstract or difficult concepts are often expressed symbolically through the use of metaphors. Four metaphors that commonly appear in rock art, in verbal descriptions of vision quests, and sometimes in ceremonial performances, are death/killing, going underwater/drowning, flight, and sexual intercourse. Let's consider each of these.

Death/Killing: A common metaphor used by a shaman to describe a trance is death or killing. The underlying logic of this metaphor is the physiological analogy between dying and entering a trance. In either case, the individual loses control of motor functions and falls down, vital signs (breath, pulse) greatly diminish or disappear, rigidity and convulsions may occur, the eyes may roll back, and the person may begin to bleed from the nose and/or mouth. A shaman who had entered a trance, accordingly, was said to have "died"; nosebleeds in shamans or nonshamans were believed to have supernatural causes, typically due to an attempt at soul theft; and epileptics experiencing convulsions were believed to be trancing.

Following this analogy, a motif portraying a dead animal can be understood as a metaphor for the shaman entering the supernatural. Since the shaman and his spirit helper were one and the same, metaphysically speaking, when the shaman entered the supernatural he died. His entry into a trance was then expressed metaphorically as a form of auto-sacrifice.

A portrayal of a human figure shooting an animal cannot be understood simply as a hunter shooting game. Instead, as expressed in an important, if at first glance enigmatic, statement by anthropologist Isabel Kelly in 1936: "It is said that rain falls when a mountain sheep is killed. Because of this some mountain sheep dreamers thought they were rain doctors." In this specific case, rain was believed to fall when a Bighorn Sheep Shaman entered the supernatural, or "killed" himself, to control the weather; such scenes of *killed* or *hunted* bighorn sheep petroglyphs are common among the engravings of the Coso Range in the western Great Basin, an area renowned for its rain shamans.

Going Underwater/Drowning: A second metaphor symbolizing a trance, particularly common in south-central California, was going underwater or drowning (combining *going underwater* with *death*). Again, the logic of this metaphor was based on the physical similarities between being in a trance and swimming, both of which combined restricted bodily movements with a sense of weightlessness, blurred vision, altered hearing, and so on. As previously noted, permanent water sources were believed to be inhabited by supernatural spirits, and some descriptions of visions say the shaman entered a pool at the start of his trance or emerged from one at its end, though—tellingly—"without getting wet."

The underwater metaphor is typically expressed in rock art by depictions of aquatic species such as frogs, water-striders, salamanders, and turtles, or, less frequently, by kelp or fish. The appearance of such images does not necessarily represent a shaman's spirit helpers; instead, it may indicate that the visionary experience was expressed using this aquatic metaphor.

Flight: As previously noted in the discussion about whirlwinds, shamans were believed to be capable of flight while in the supernatural. Several altered state sensations contribute to this perception, including a sense of weightlessness, a dissociative mental state that can produce an "out-of-body" experience, and changes in vision that make objects in the real world appear as if they are at a great distance.

The centrality of the flight metaphor in Native American cultures is best expressed in shamanistic costumery: almost invariably, shamans' ritual clothing included bird parts. In far-western North America, ritual clothing included feathered headdresses, skirts or kilts, shawls or robes, and feathered bandoliers or strands of eagle down. The flight metaphor is expressed in rock art in a number of similar ways, with shamans sometimes wearing avian costumes or appearing partially transformed into a bird, usually having clawlike bird feet and/or feathered wings. They also appear with concentric circles (whirlwind) representing their heads or with pinwheels above their heads. Less commonly, birds themselves are sometimes depicted.

Sexual Intercourse: The final metaphor symbolized in far-western North American rock art is sexual intercourse. Some of the hallucinogens shamans used to enter an altered state, including jimsonweed and native tobacco, resulted in sexual arousal. There is an obvious physical analogy between the caves and rock shelters typical of many rock art sites and vaginas/wombs. Accordingly, shamans sometimes expressed their supernatural experiences in terms of sexual intercourse with a supernatural spirit, and some religious ceremonies involved simulated coitus. In myths, rattlesnakes guard the vaginas of women; at rock art sites, they guard entrances to the supernatural.

The expression of sexual intercourse as a metaphor in far-western North American rock art was seldom graphic, in the literal sense of the term. Depictions of intercourse are common in the rock art of other regions, such as the Southwest, but they are rare in the Far West. In southern California and Nevada, this metaphor is expressed typically by vulva-shaped motifs, indicating that the rock art site itself symbolizes the vagina. The shaman's entry through his portal to the sacred realm was thus a kind of ritual intercourse with his rock art site, a metaphorical vagina.

Puberty Initiates and Rock Art

Our discussion so far has emphasized the role shamans played in rock art of the Far West. In southwestern California and the Colorado River region, puberty initiates also made rock art. They entered the sacred realm during their coming-of-age ritual and depicted the spirit helpers they saw in their visions. The neuropsychological model and the symbolism discussed above apply equally to the art of initiates and shamans; both forms are fundamentally shamanistic in origin, nature, intent, and symbolism.

In southwestern California elaborate group rituals were held separately for boys and girls at about the time they came of age. Intended to prepare the youngsters for adulthood, the ceremonies involved isolation from the village, instructions in sacred beliefs, dances, songs, lessons in expected adult behavior, and a visionary experience to obtain a spirit helper. While shamans obtained supernatural power through their spirit helpers, nonshamans obtained helpers to help them live successful and happy lives. The principal difference between a shaman and a nonshaman with reference to spirit helpers and supernatural power was one of degree, not kind. Shamans often had several helpers with whom they had particularly intense and intimate relationships. Nonshamans, in contrast, whether male or female, might have only one helper for a variety of everyday pursuits—from imparting good luck in hunting or gambling for a man to easing the pain of childbirth for a woman.

The girls' puberty ceremony in southwestern California—last conducted in northern San Diego County in the 1890s—is particularly well known. The girls were isolated for a number of days, ritually enacting the procedures and carefully following the taboos of a woman during childbirth. The isolation, sensory deprivation, and fasting the girls experienced could result in hallucinations, but to ensure that end the girls also ingested tobacco. At the culmination of the ceremony, the girls ran a ritual race to a special rock—

known as "Paha's House"—near their village. Paha is the term for shaman, so the race ended at the shaman's house. The race demonstrated the girls' health and vitality, with the winner considered the healthiest and predicted to be the longest lived. In fact, the winner of the last girls' puberty race of the 1890s lived longer than any of her ritual cohorts.

The high point of the initiation, underscoring its great symbolic importance, was the girls' arrival at Paha's House to paint the spirit helpers they saw during their visions. In the majority of cases, the girls' spirit helper was a rattlesnake, a symbol closely associated with women throughout the Far West. As noted above, rattlesnakes guard the vaginas of mythic women just as they guard entrances to the supernatural at rock art sites. The puberty initiation was not considered successful unless a "supernatural snake who lived in a cave somewhere" appeared at the ceremony and showed itself to the girls during their visions, thus becoming their spirit helper. Zigzags and diamond-chain motifs—symbolizing rattlesnakes throughout the Far West—heavily dominate the girls' art. Other motifs, especially entoptic patterns, also appear, as do some stick-figure humans, birds, and other forms. Hand prints symbolize that the girls had *touched* the supernatural.

The girls' art is almost invariably painted in red, the color most frequently associated with females. Although male shamans could also paint in red, its dominant use in girls' puberty art ties in with their menstrual bleeding and the "female" direction, west, which bears the color of the sunset. Following this symbolic logic, black was most often associated with males and their direction, east, toward the night. It is also worth noting that the word *Paha*, meaning shaman, is another name for the California red racer, a snake that experiences both black and red color phases during its life. In southwestern California, it was (incorrectly) believed that the red and black phases of these snakes were actually the contrasting sexes. The shaman/Paha, then, painted himself bilaterally black and red for rituals, thus, like the red racer, conjoining or mediating the male and female principles of the world.

We know considerably less about boys' ritual art in southwestern California than we do about the girls', primarily because the boys' initiation ceremonies ceased considerably earlier. There are similarities, however, such as painting rock art following a ritual race to conclude the initiation; and the ritual events leading up to that event are well described. In addition to isolation and physical exertion, the boys also ingested a hallucinogen — but they took jimsonweed rather than tobacco, requiring carefully controlled circumstances due to its strong and unpredictable effects. During the boys' isolation from the village, shamans created sand paintings on the ground that symbolized their cosmological beliefs and facilitated the instructions of the youngsters. But, like the girls' ceremony, a primary goal of the boys' initiation was the acquisition of a spirit helper during a vision.

Existing ethnographic descriptions of the boys' art are extremely brief, and we have no specific information linking any known rock art site to a boys' ceremony. In one description, the boys appear to have painted on the same rock as the girls, although on a different side. We are told that the boys painted in black, befitting the red/black, female/male color symbolism discussed above, but we have very few descriptions of the boys' motifs, which include groups of circles and nested curves (both entoptic patterns), and a sticklike human figure. A verbal account also indicates that at least one boy's painting, rendered as a "streak," presumably a line, represented Wanawut, a male humanlike spirit personified as the Milky Way. In other segments of the boys' ceremony, a fetish image of Wanawut, woven out of reeds, was displayed. In this case, Wanawut took the form of a human figure.

Our ethnographic knowledge of initiatory art among the Colorado River groups is also limited. But we know the boys made rock art in their puberty initiations, during which their nasal septums were pierced for nose ornaments and they endured extreme physical stress, fasting, and sleep deprivation (again, conditions that result in an altered state). Following the general pattern for the Colorado River region, supernatural power among these groups was linked to mythic time-space, so the hallucinations the initiates saw and the images derived from them were connected with the reexperiencing of their mythological past. Although the puberty initiation sites differed from the shamans' vision quest sites, there are indications that the boys visited shamans' rock art sites to examine and study the images prior to making their own elsewhere.

Furthermore, during major life events such as marriage, becoming a parent, the death of a loved one, and so on, a Native American from the Colorado River region might revitalize his supernatural power with subsequent vision quests and petroglyph creation.

Based on cases in southwestern California and the Colorado River regions, we know that not all rock art in the Far West was made by shamans. Yet the art and the various rituals behind it were shamanistic in nature. In regions where supernatural power was widely maintained in society—and where the distinction between nonshamans and shamans, as ritual specialists, was less finely drawn than in other parts of the Far West—most or all members of a society might make rock art. Still, it was created in a shamanistic context that involved vision questing, the acquisition of supernatural power, and the depiction of hallucinatory visions.

OTHER INTERPRETATIONS

All the information presented so far points to a unitary explanation for the rock art of the Far West. But can we be sure none of the art resulted from other traditions or rituals, or symbolized things other than visions of the supernatural? Although there are still some unknowns concerning this art, the large majority may be explained by this interpretation in spite of numerous contrasting interpretations by archaeologists, particularly in the older literature. After covering a few general points about the issue, we'll look at why those older views are probably incorrect.

First, the existing ethnographic record of Native American lifeways, as collected by anthropologists from elderly informants in the early 1900s, makes no mention of other origins or functions for the art; the record is unanimous and unequivocal. Many of the older interpretations were predicated on the belief that there was no ethnographic record about the creation of rock art—a circumstance we now know is false. Instead, the older views relied on vague and misguided notions about the "primitive" mentality of Native Americans, and they drew superficial and incorrect analogies between the far-western groups and other native groups. Since we now know that detailed ethnography exists and that it was derived precisely from the people who made the art, any interpretations that fail to consider this valuable resource are necessarily shortsighted. Simply put, it is unduly wrongheaded to deny the Native American voice in the interpretation of this art.

Second, archaeologists recently have begun to interpret rock art in other regions by examining ethnographic records across the hemisphere, much as I have done for the Far West. This research increasingly tells us that much of the rock art made by hunter-gatherer groups—including those on the Columbia Plateau, the northern and southern plains, the eastern woodlands, and even some lowland South American groups—resulted from shamanistic practices similar to those I have described above. This interpretation of the art is part of a widespread pattern that may represent a tradition very ancient and fundamental to Native American cultures.

Third, my discussion has emphasized what might be considered the primary purpose and origin of the art. Certainly the art may have served other purposes as well. For example, the native peoples recognized particular rock art sites in the Great Basin as locations of power. Nonshamans sometimes visited them for what might best be called "folk cures," during which they would pray, fast, and leave offerings in the hopes of relieving some ailment. So, even within the context of a shamanistic/vision quest interpretation of the art, the potential for other secondary uses of the art must also be acknowledged.

Fourth, there are certain specific examples of rock art in California that do not fit the ethnographic pattern outlined above. Two of these, discussed in reference to sites 20 and 38 subsequently, are the the earth figures or intaglios of the Colorado River region and the so-called maze-style art of western Riverside and northern San Diego Counties. While we have ethnographic information on the intaglios, indicating they were used in ceremonies associated with specific myths, the origin and function of the maze-style art remains a scientific mystery.

We have much yet to learn about far-western North American rock art, even at the most basic level, and we may ultimately discover some alternative meanings. Further, the interpretations offered here pertain specifically to the more recent examples of rock art and are derived entirely from historical ethnographies. Although it is likely that the historical traditions underlying rock art also apply to distant prehistory, that hypothesis is not yet proven.

Cultures often change over time, and it is possible that the meanings and purposes of rock art have as well. Still, our best hypothesis at this point is that the very early art was made for reasons similar, albeit not necessarily identical, to those during the more recent Native American past.

Hunting Magic

One early interpretation of rock art attributed it to "hunting magic." Although never clearly explained, the underlying assumption seemed to be that Native Americans believed painting or engraving the image of an animal on rocks would somehow (magically) cause that animal to appear or to reproduce successfully, or perhaps cause it to fall easily from the hunter's arrows. The interpretation further contended that hunting-magic sites occurred on game trails, and that this locational association "proved" the hypothesis.

There are four flaws with this interpretation. First, there is no evidence that Native American groups in California or Nevada believed in this type of hunting magic, and none claimed to have made the art for such a purpose. Early archaeologists ignored this problem because they knew of no ethnography relating to the art and therefore inferred that it was solely prehistoric in age.

Second, an examination of the art in regions where hunting magic supposedly occurred reveals that relatively little of it actually represents the game that hunters might have "created" by making the art. For example, less than 10 percent of the petroglyphs in Nevada represent game animals that hunting magic purportedly sought to conjure up; the vast majority of the motifs are, in fact, entoptic patterns.

Third, almost all the rock art sites said to lie on game trails are actually located at village sites near permanent springs. The association between art and game trails is therefore incorrect.

Finally, at a more pernicious level, this type of "explanation" reduces the religious, symbolic, and metaphysical beliefs of Native Americans to no more than a concern with food and subsistence. Yet we know that Native Americans had—and, where possible, continue to maintain—intellectual and symbolic traditions as rich and complex as our own.

Boundary/Route Markers

Another early interpretation of rock art suggests that it marked territorial boundaries, guided travelers along trails, and so on. Again, there is no ethnographic support or other evidence that rock art served such a purpose. As the anthropologist Alfred Kroeber noted in 1925, Native Americans knew their regional geography so well that they did not need signs to guide them, and, in terms of territorial markers, their inclination would have been to hide boundaries rather than advertise them. As a more or less permanent feature on the landscape, however, the art probably served as a geographical referent on occasion, just as we might tell a lost driver to "turn right at the red barn"; in such a case, the building serves a useful secondary purpose as a geographical referent, but that does not tell us the primary reason for its construction. Similarly, rock art was not made to serve as directional signs.

Solstice Observatories

Some researchers believe rock art sites were solstice observatories, with much of the art portraying celestial phenomena. This interpretation has become so popular that a major museum in southern California includes it in an exhibit to explain rock art. Popularity aside, the support for this interpretation is at best equivocal, and it therefore warrants discussion in some detail.

The solstice observatory hypothesis is not necessarily incompatible with the shamanistic interpretation presented in this guide. Shamans could have selected vision quest sites based on observing the solstice, among other criteria. Furthermore, we know that far-western North American peoples had detailed and sophisticated knowledge of astronomy; the winter solstice was an important ritual period for them, and they knew when to expect it by tracking the sun. We also have an ethnographic statement claiming that two Chumash shamans painted rock art essentially at the time of the solstice. At a couple of sites, in addition, archaeologists have observed curious lighting effects on the morning of the solstice, such as a dagger-shaped shaft of light crossing a central motif on a painted panel. Again, however, this hypothesis was developed at a time when archaeologists knew little about the existence of ethnography on the art.

The solstice observatory hypothesis, then, faces several problems. First, we now recognize that a wealth of ethnographic data exists to tell us about the origin, function, and meaning of rock art in the Far West. With the exception of the single statement (noted in the previous paragraph) about Chumash shamans painting at the time of the solstice, which is significant, none of the ethnography ties the art to the solstice or to solstice observations; other ethnographic accounts make no mention of rock art sites used for such purposes. The negative evidence, which is substantial, thus does not support the solstice observatory interpretation, in spite of the importance of that day to Native Americans in the Far West. For example, one of their most important and most elaborate religious rituals—the Mourning Ceremony—was conducted during the solstice; but its purpose was to commemorate the dead, not to observe the solstice. The fact that Chumash shamans created rock art at this time could reflect any number of purposes other than to observe the solstice.

Second, the enthusiasm of rock art researchers to relate the few instances of seemingly nonrandom lighting events to the solstice overwhelms their scientific objectivity. The effort of reaching a remote site before dawn in the middle of winter to watch the sunrise can lead people to interpret meaning into their observations when none exists. Contrary to many contentions, the simple fact that the sun can be seen rising over mountains, hills, or peaks on the morning of the solstice, as it does on every other day of the year, or that the sunlight illuminates a panel of rock art on the solstice, as it does on many other days of the year, is not evidence that the sites served as observatories. When viewed critically, in fact, the alleged empirical evidence for most of the sites suspected of serving as observatories seems more likely to disprove the theory.

Third, there is a danger of ethnocentrism when we attempt to understand Native American religion, rituals, art, and symbolism on the basis of their equivalence to western European science. We value science for many good reasons; some may even think of science as the modern religion for our culture. But it is a fallacy to predicate our interest in Native Americans on how closely their beliefs and intellectual accomplishments match our own scientific achievements, as our obsession with proving them to have been astronomers seems intent on doing. It is in fact the differences, not the similarities, between Native American cultures and our own that are most interesting. These cultures deserve to be understood in their own rights and contexts, not simply as reflections of ours. As individuals, we may choose to replace religion with science, but we must question the propriety of projecting such a replacement on Native American cultures.

There is a shred of support for the hypothesis that a few rock art sites may have served as solstice observatories. A few of the sites indeed experience unusual and nonrandom lighting events during the winter solstice. It is also likely that some rock art motifs portray celestial phenomena; for instance, certain supernatural spirits seen by shamans and initiates in their visions were personified planets and stars. But the notion that most or many of the sites served as solstice observatories, or that clusters of geometric motifs were necessarily star maps, is not supported in the ethnographic record or at the sites themselves.

MAKING THE ART

While the ethnographic record provides us a coherent interpretation of rock art in the Far West, the three *types* of rock art warrant a brief description. Although occasionally found on the coast, *petroglyphs*, or rock engravings, are most common in the desert portions of the Far West, where time and weather produce a dark coating of "rock (or desert) varnish" on many boulders and cliff faces. Native Americans made petroglyphs by pecking or scratching through the rock varnish with stone tools such as hammerstones, abraders, and lithic knives and chisels to expose the lighter-colored rock beneath. The sharp contrast between the light-colored etched portion and the dark surrounding surface highlights the image. Replicative experiments using stone tools indicate that an average petroglyph can be engraved in about an hour or two.

Pictographs, or rock paintings, while most common in the coastal provinces, the peninsular mountains, and the Sierra Nevada, are also found in the desert regions. The Native Americans made red, white, and black paint by mixing mineral earths—usually hematite for red, kaolin clay for white, and charcoal for black—with organic binders such as animal blood, rendered fat, or oils from crushed seeds. In the Great Basin, they often obtained mineral earths from hot springs, which they believed were inhabited by supernatural spirits, thus imbuing the paint with supernatural potency. Native groups in southwestern California made red pigment by collecting and drying a particular type of pond algae. They ground the pigments with bedrock mortars commonly found near pictograph sites and applied the paint to rocks with brushes made from vegetable fibers or animal hairs, or they daubed it on with their fingers or hands.

Natives in the desert regions of the Far West also made earth figures called *intaglios* or *geoglyphs*. These are especially common along the lower Colorado River, indicating their particular importance to Yuman speakers, but they are also present in the Mojave Desert. Earth figures are large, monumental images fully akin to the better-known "Nazca Lines" of Peru. They were made in two ways. On the terraces above the Colorado River, where the ground is covered by darkly varnished rocks and cobbles forming a "desert pavement," the earth figures were created by scraping aside this pavement to expose the lighter soil below. In the Great Basin, where desert pavement is less common, earth figures were typically created by aligning cobbles and small boulders into the desired image.

DATING THE ART

One of the most common questions people ask about the rock art of far-western North America is: How old are the images? Based on the ethnographic data collected by anthropologists from older Native Americans in the first few decades of the twentieth century, we know the cultural tradition of creating rock art continued at least into the nineteenth century, and perhaps beyond. We have no evidence that such a tradition still exists, but some Native Americans maintain the knowledge and occasionally discuss it. Until recently, archaeologists assumed that much if not all of the art predated the historical period; some modern books and articles on far-western North American rock art still express this belief. But the evidence, including the ethnographic record, clearly shows this is not the case. The fact that some of the art dates to the last few hundred years partly answers the question of its age. We can assume that some of the art must be somewhat older, but how much older?

Archaeologists can draw inferences about the age of rock art three ways. First, they look for time-specific themes or subject matter in the images. Rock art depicting horses and riders, for example, or human figures wearing Euro-American hats, confirms that some of the art was made after European "discovery" of the New World; such motifs must postdate the initial appearance of Europeans in western North America in the sixteenth century, and they may postdate widespread cultural contact in the late eighteenth and early nineteenth centuries. Art depicting horses and riders exists at numerous sites in south-central California, the Great Basin, and the Colorado River region.

Time-specific subject matter can provide clues about older art in a few cases. Throughout the Far West, bow-and-arrow technology appeared about A.D. 500, replacing atlatls (spear-throwing boards) as the primary means of flinging projectiles through the air. So bow-and-arrow motifs must necessarily be less than 1,500 years old while depictions of atlatls should be older than 1,500 years. Similarly, a few rock art sites in western North America depict what appear to be extinct Pleistocene animals such as the Imperial Mammoth or the Ice Age Camelops; if this identification is correct, it should prove that some of the art may go back 10,000 years or more.

The second clue archaeologists use to determine the antiquity of art is its condition relative to other art at a particular site. Simply stated, a more deteriorated and/or faint painting should be older than one that shows less deterioration and is brighter; or a darker, more heavily revarnished engraving should be older than one that is lighter and less revarnished. But archaeologists must be careful when considering microenvironmental and other factors that can affect the condition of art at any particular site; art directly exposed to weather, for instance, cannot be compared fairly with art that has been protected from such conditions, even though the depictions may lie relatively near to each other at the site.

While using relative condition to determine the age of rock art depends on a variety of factors, archaeologists can draw helpful inferences through this method. By comparing older and newer photographs, a number of researchers have noticed visible deterioration of some pictograph sites in south-central and southwestern California in the last 50 to 100 years. Although the rate of deterioration may have accelerated recently due to acid rain, smog, and other environmental pollutants, it appears that much of the art is rapidly eroding, suggesting that it may be less than 1,000 years old.

In the Great Basin, archaeologists have cross-checked relative condition with subject matter to achieve slightly stronger inferences about the age of rock art. For example, atlatl motifs in the Coso Range, which should date to 1,500 years or older, typically exhibit moderate to heavy revarnishing; bow-and-arrow engravings, which are less than 1,500 years old, generally have little or no obvious revarnishing. When these factors are combined and compared to the Coso petroglyphs as a whole, it can be reasonably inferred that a majority, though by no means all, of the engravings in this region were made within the last 1,500 to 1,000 years.

Ideally, of course, we would like to determine specific dates for particular paintings and engravings. This is possible by using chronometric techniques, the third method by which archaeologists can judge the antiquity of rock art. Chronometric techniques have only been developed since the early 1980s; they are still experimental, and their results must be viewed as provisional. This method of dating rock art is very expensive to perform with a typical archaeological research budget, so it has not yet yielded a large body of dates. Nonetheless, much of the art that has been chronometrically dated, in a research project begun in 1981 by Arizona State University geomorphologist Ronald Dorn and me, is in the Mojave Desert portion of the Great Basin. Our efforts so far have been primarily devoted to dating the rock varnish that accumulates over time on petroglyphs.

The age of this rock varnish can be determined in three ways. The first way calls for radiocarbon dating the small quantities of organic matter, such as lichens or pollen, that get trapped in the engraved area of a petroglyph as the rock varnish accumulates on top of them. We do this with a nuclear accelerator, in a manner similar to that used for pictographs (discussed below). The second way measures the proportions of major inorganic trace elements in the varnish, some of which leach out at a known rate over time while others remain more stable. Using an electron microprobe analysis, or measuring the particle-induced x-ray emissions (PIXE), yields what we call "cation-ratio dates." Because radiocarbon dating and cation-ratio dating involve analyses of different phenomena using different techniques (the nuclear decay of organic matter, and the relative chemical change in rock varnish inorganic constituents, respectively), they are independent techniques and therefore can be used to cross-check results.

The third way, viewing the micromorphology of rock varnish under a scanning electron microscope (SEM), has also proven to exhibit different characteristic forms, depending upon when the varnish began to develop. During the Pleistocene or Ice Age, or before about 11,000 years ago, environmental conditions were wetter and less dusty in the West. Varnish that developed during these conditions exhibits a rounded, lumpy shape called "botryoidal" when viewed in cross-section under an SEM. Rock varnish that formed in the drier, dustier conditions of the last 11,000 years, in contrast, has a flat, platy structure called "lamellate" when a cross-section is viewed microscopically. Typically, a petroglyph engraved more than 11,000 years ago shows botryoidal layers of varnish overlain with layers of lamellate. A petroglyph engraved less than 11,000 years ago has only a lamellate varnish. We can use this change in microstructure as another independent check on our radiocarbon and cation-ratio dates.

We can also use these three varnish-dating techniques to date the earth figures along the Colorado River. As natural geomorphological processes push small rocks from their embedded position in the soil onto the recessed surface of the earth figures, the rocks eventually develop a coating of varnish. We know this coating of rock varnish postdates the creation of the earth figure, so dating the varnish provides us with what we term a "minimum age" for the art itself, which must be older than our calculated dates on the naturally upthrusted cobbles within the figures.

For pictographs, chronometric analysis is limited to radiocarbon dating minute amounts of organic matter in the paint. The organic matter may come from either the pigment binder (blood, tallow, or seed oil), or, in the case of black paint, the pigment it-self—charcoal. Because the amount of organic material present in the paint is so small, we must use a nuclear accelerator to calculate the amount of radioactive decay that has occurred in the sample since it ceased to be a living organism (such as a tree for the charcoal or an animal for blood or tallow). Although radiocarbon dating is well established as an archaeological technique, the use of a nuclear accelerator to determine ages from very small samples has only been possible since the mid-1980s, and the application of accelerator radiocarbon dating to pictographs is still new.

We have used these different techniques to obtain preliminary chronometric dates on a series of pictographs, petroglyphs, and earth figures. The ethnographic record, time-specific themes or subject matter, and, in some cases, relative condition all demonstrate that some if not much of the art in the Far West is, archaeologically speaking, relatively recent—ranging from 2,000 to 100 years or less in age. These relative dates are confirmed by our chronometric assays, a number of which indicate art made within the last 500 years. But dates on geoglyphs along the Colorado River indicate that at least some of the earth figures are almost 3,000 years old. And an experimental radiocarbon date on a pictograph from the Mojave Desert indicates it is 9,000 years old. Perhaps most surprising are the varnish dates we obtained from six Coso Range petroglyphs, showing ages from 12,900 to 19,100 years old. These dates are admittedly controversial among archaeologists, but if they prove correct with subsequent testing they will indicate that some of the rock art of far-western North America may be as old as the famous Paleolithic cave paintings of Lascaux, France, and Altamira, Spain.

Although the chronometric dating of rock art is still in the experimental stage, it suggests that some far-western North American rock art may be very ancient, even while the ethnographic record and other evidence indicates that some of the art continued to be made into historical times. All of this evidence suggests one conclusion: The production of rock art was a tradition most likely brought to North America by the first Native American migrants during the initial peopling of the continent, and it was practiced by them for as long as their traditional lifeways were followed. It is not surprising that rock painting and engraving are essentially universal practices among hunter-gatherer groups worldwide.

I think it only reasonable to conclude that if Native Americans were present in a given region, they likely made rock art. The age of art in any specific area is largely a function of the antiquity and duration of Native American occupation of that region, albeit over time much of the earlier art can be expected to have eroded away. The majority of far-western North American rock art, accordingly, is probably relatively recent—less than 2,000 years in age—simply due to the limits of natural preservation. Some of the art may be quite ancient, however, especially in the desert regions where drier conditions enhance its preservation.

Early Euro-American settlers widely derided Native Americans of the Far West, viewing their societies and cultures as primitive and their technologies as childlike. But because the Far West has a Mediterranean-type climate, with rainfall limited to the winter months, agriculture (based on the Western Hemisphere's summer-growing crops) could not spread into the region in any significant degree, constraining the potential for population growth. Further, the nomadic lifestyle of the far-western Native Americans countered their potential to accumulate significant material wealth and large inventories of tools. Yet more recently, these same cultural patterns—relatively low population density, a hunting and gathering way of life, and a limited range of material items—have led many Euro-Americans, including archaeologists, to categorize these same Native American cultures as "simple," "non-complex," and "representing the starting point for human social evolution."

An understanding of the Native American rock art of the Far West, however, provides a counterpoint to and belies such beliefs. Certainly this art reflects artistic and symbolic systems, and metaphysical and religious beliefs and practices, as complex as those we maintain in our own highly technological society. Native American rock art was simply one manifestation—in this case, one intended to portray visions of the supernatural world—of a complex and sophisticated worldview that we would benefit from paying heed to, even today.

APPRECIATING ROCK ART SITES

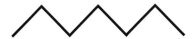

The rock art sites of southern California and southern Nevada, as should now be clear, represent different things to different groups in our society. To modern Native Americans, descendants of the original artists, they are sacred locales fully akin to our churches and places of worship. To archaeologists, the sites represent important yet fragile sources of information about prehistoric cultures and beliefs. And to American society at large, rock art sites constitute part of our communal cultural heritage, a heritage that in this case is valuable for its combined historical, aesthetic, and religious characteristics. Keep each of these perspectives in mind when you visit the sites; they all deserve our full accommodation and respect. More to the point, we have a responsibility to ensure that our visits to the sites do nothing to diminish their sacredness or their value for scientific study. We must help preserve these sites so that future generations may also appreciate them.

PRESERVING THE PAST

It is reasonable to assume that any reader of a book such as this would abhor overt vandalism—such as painting graffiti, stealing or damaging panels, or leaving trash—at rock art sites, and there is no need to belabor this topic here. More important then, and requiring some discussion, is the fact that our visits to sites can inadvertently contribute to their degradation, especially when we recognize that our individual acts are amplified by the multitude of visitors to the sites. But with care, and a little forethought, we can eliminate or greatly diminish our impact on these fragile cultural resources. Moreover, by visiting the sites, especially the more remote and less frequented ones, we can even have a positive effect on preservation. Our mere presence will discourage all but the most determined vandals. Further, our few words of caution can educate the unaware about the potentially deleterious impacts of their behavior. So I encourage you to consider how visiting these sites can favorably affect their preservation.

One of the most important guidelines to remember while visiting the sites is that the panels of art are often only portions of larger archaeological locales. The sites sometimes lie within villages, and thus are found in association with artifacts, prehistoric refuse deposits, and so on. Because archaeological research depends heavily on relating the kinds of artifacts and features that may be found together on a site, it is crucial that you do not touch or disturb any artifacts you see. The location of an artifact is extremely important to archaeologists. While it may be tempting to pick up and pass around an arrowhead, shell bead, or other artifact, this seemingly innocuous behavior compromises the integrity of a site. And theft of any artifact not only diminishes the site's integrity but also is prohibited by law.

Another guiding principle for visiting rock art sites is to remember that the art—whether paintings, engravings, or earth figures—is inherently fragile and can be destroyed all too easily. Brushing up against a pictograph may knock off small or sometimes large fragments of the paint and may even contribute to detaching slabs of the underlying rock off the panel. Touching the art with your fingers can also knock off the paint, and it leaves an invisible residue of skin oils on the art that attracts dirt and may destroy the potential for chronometric dating. So watch where you place your hands when visiting the sites, and be careful that your backpack, hat, shoulders, and so on do not rub against the panels of art.

Not so long ago it was a common practice, even among archaeologists, to spray water on pictographs, or chalk in the recesses of petroglyphs, to enhance photographs of the art. Visitors also made rubbings and castings of petroglyphs to record the art. We now know that such practices must be avoided at all costs. Applying water greatly accelerates the erosion of ancient paints, while chalking, rubbing, and casting destroy the potential for chronometric dating of engravings. Chalking in an outline of a petroglyph also greatly diminishes its aesthetic qualities, often ruining the experience for subsequent visitors. It is disheartening to see a figure that has been sloppily or incorrectly chalked, creating an image entirely unfaithful to the original.

Another major factor that degrades rock art sites is the accumulation of dust on the art, which may occur even when the paintings or engravings are on vertical panels or ceilings. Much of the dust is wind-borne and thus attributable to natural processes, but the dust generated by "visitor pressure"—the shuffling of countless numbers of human feet by a site—also takes a heavy toll. In certain cases, the resource managers of rock art sites have placed stone pavers or wooden walkways around sites to control dust, but in many cases these protective measures are impractical. We may not be able to entirely eliminate our creation of dust, but by being aware of the problem we can minimize its impact on the art by treading lightly.

Rock art conservators and site managers have conducted studies of visitor behavior at sites to better understand the impacts of human visitation on rock art. One conclusion derived from these studies is that children pose the greatest hazard to the sites. They tend to be more physically active than adults, more prone to touch and rub against the art, more likely to pick up artifacts from the ground, and so on. Adults must exercise caution in taking children to rock art sites and keep the safety of the art well in mind. This is not meant to discourage taking youngsters to rock art sites; my own daughter has accompanied me to countless sites around the world since she was only six weeks old, and she has visited all of the sites in this guide. Instead, the future well-being of rock art sites depends entirely upon the degree to which we teach our children the importance of caring for this valuable heritage; education truly is our best long-term strategy for site preservation. I encourage parents to take their children to visit these sites, but they must accept the added responsibility of educating the children and monitoring their behavior.

One final point about visitor behavior at rock art sites must not be overlooked. Remember that these sites were, and in many cases still are, sacred places for Native Americans. Any behavior that would be inappropriate in a church or at a cemetery should not be practiced at the sites. Rock art sites deserve respect as sacred places.

PHOTOGRAPHING ROCK ART

Many rock art site visitors quickly become avid photographers. Rock art photography is sufficiently technical to warrant a manual of its own, but, with a little work and patience, anyone can take reasonably good photos of rock art. There are a few brief introductory principles about photographing rock art that can help place you on the road to better rock art photography.

Two common conditions that make rock art difficult to photograph are poor light and faint or indistinct art. Past methods of enhancing faded or indistinct art, such as wetting or chalking the motifs, must be avoided in all cases; in other words, never physically touch the art with anything. Lighting, however, can often be controlled to produce contrast between the art and its background.

Perhaps the most common lighting problem, especially in caves and rock shelters, is simply low light. This can be overcome in several simple ways, the most obvious of which are the use of higher-speed film and/or a tripod. Although I commonly use ASA 100 film when I visit rock art sites, I generally carry a roll or two of ASA 200 or 400 with me as the easiest solution to unforeseen low-lighting problems. When rock art panels or motifs are in a shadow, you can throw additional light on them with a reflector. Low-angle reflected light can be particularly effective when photographing petroglyphs because the "raking" effect of the light casts a shadow into the engravings, thus improving their visual contrast with the surrounding rock panel. There are a number of lightweight, collapsible cloth reflectors for sale at photo supply stores, but an inexpensive, easily transportable alternative is to use an aluminum-colored "space blanket," available at any sporting goods store.

Although the use of flashes is often prohibited at art museums, it is a myth that modern flash equipment damages paintings or pigment. Using a flash at rock art sites, accordingly, will not harm the art and, especially at pictograph sites, can yield very good photos; in particular, the flash often enhances the brightness of red pigments. While flash photography may not always yield colors entirely true to the original, the resulting photograph can be valuable for study purposes and is aesthetically pleasing. The best flash results can usually be obtained by using a flash cable about eight feet long, positioning the flash to the left of the camera, and pointing it toward the art at a 45-degree angle. This brings out the texture of the rock surface as well as illuminating the art itself.

Photographing rock engravings, often located on large open-air cliff faces, presents a slightly different lighting problem: the petroglyphs may appear washed out from too much light and too little contrast. The lighting varies not only with time of day but also according to the season. A polarizing filter can help control the contrast, but patience and planning your visits to take advantage of the best lighting conditions are often the key to good petroglyph photos.

A final consideration in rock art photography is film and processing. Certain brands and types of film, as well as processing techniques, emphasize different segments of the color spectrum. For example, some intentionally brighten the colors that may transform a sky from light blue to deep blue while others may provide truer color or emphasize the red end of the color spectrum, such as tungsten or indoor films. If your goal is an exact reproduction of the rock art image as it currently exists, use a color card and true-color film and processing. Red is a common color in much of the painted art in the Far West. Since I use photographs more for study purposes than as exact reproductions of the art, I sometimes prefer film and processing that enhances the reds, making the images more visible even though they are less true to reality. Consult the experts at your local camera shop for specific advice but, in general, expect only a moderate gain when using high-priced professional film, which requires rigorous temperature-controlled handling. Most brands of over-the-counter or "amateur" slide and print film are adequate for exceptional rock art photos.

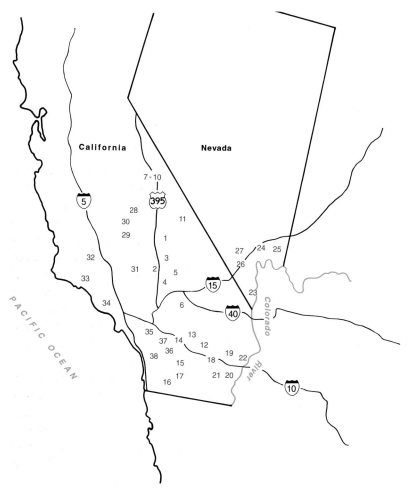

Map 2: Site locations.

VISITING THE SITES

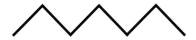

The rock art sites selected for this guide (see map 2) are open to the public and generally have ongoing management and preservation programs in place. All the sites are accessible by vehicle except one—site 17, McCoy Spring—which requires a substantial hike. Unless noted otherwise, most of the sites allow unrestricted visitation during daylight hours. Some sites, especially those in the Mojave and Colorado Deserts, lie in relatively remote areas. Take normal precautions when traveling in the desert—carry extra water and a shovel, and make sure your car is in good repair. In most cases, a standard two-wheel-drive sedan is adequate to reach the sites; those requiring four-wheel drive—typically the more remote desert sites—are explicitly noted, but even these do not require unusual off-road skills or derring-do to reach.

The maps in this guide provide detailed directions to each site. But the distances shown, even though given in tenths of a mile, are approximate. This is partly due to the inherent crudeness of automobile mileage odometers and partly due to the fact that desert roads often include several options among parallel tracks and curves; distances can easily vary by a few tenths of a mile. For the more remote and harder-to-find sites, I have included references to the appropriate U.S. Geological Survey (USGS) topographical quadrangles and/or Bureau of Land Management (BLM) Desert Access maps where appropriate. You should not *need* USGS or BLM maps to find the sites, but they offer a wider geographical frame of reference than the site maps.

All of the sites are located in rocky areas frequented by rattle-snakes, especially in warm weather, so watch where you step. Also, changing weather conditions constantly loosen rocks that may fall or become dislodged as people walk or grab them, so be careful where you put your hands and your feet, both to protect the art and to protect yourself from potential harm.

MAP 93
C-7

Mojave Desert and Owens Valley Region

SITE 1

Coso Range Petroglyphs
Ridgecrest, CA

Making Rain in the Desert

One of the most spectacular concentrations of rock art sites in North America—if not the world—is in the Coso Range, in the eastern California portion of the Great Basin. The engravings in these mountains were made by Numic-speaking Shoshone, Northern and Southern Paiute, and Kawaiisu peoples whose territory extended from eastern California through the Great Basin to western Wyoming. Numic shamans traveled from at least as far away as Utah to conduct vision quests in the Cosos. Their petroglyphs are dominated by depictions of bighorn or mountain sheep, the special spirit helpers of Rain Shamans (see photo 1).

Shamans traveled great distances to visit the Cosos because these mountains were considered a particularly likely place to acquire the power to control weather. Some of the sheep petroglyphs are depicted as if killed or being shot by a human figure with a bow and arrow (see site 24). Such scenes refer to a Numic metaphor for rainmaking. Entering a trance, a necessary part of making rain, was metaphorically described as "dying" because of the physiological similarities—loss of consciousness, bodily collapse, diminished vital signs, occasional convulsions and bleeding from the mouth or nose, and rolled-back eyes. Since the shaman and his spirit helper, the bighorn sheep, were considered one and the same, and since the shaman's trance was a metaphoric form of ritual death, the shaman made rain by "killing a bighorn" (or killing himself, a supernatural bighorn) and entering a trance. Thus, petroglyphs depicting a human shooting a sheep have nothing to do with hunting; they are metaphoric symbols indicating that a Bighorn Shaman came to this location and entered a trance to bring rain.

49

Another point worth noting about the Coso Range bighorn petroglyphs is that they all depict sexually mature adult male sheep (according to my study of over 350 bighorn engravings), further linking adult male shamans with adult male sheep. This contradicts suggestions that female or immature sheep were connected in some way to fertility rituals. The sex of male sheep is evident sometimes by the presence of a penis sheath, a downcurving line at about mid-belly, or more commonly simply by the presence of fully curved horns. Male bighorn sheep retain juvenile bodily characteristics, including small horns, until they reach sexual maturity at about six years of age, at which time they develop the large incurving horns most commonly portrayed in the petroglyphs; females never develop such pronounced horns.

Other common motifs in the Cosos petroglyphs include geometric designs, the entoptic patterns shamans saw during their visions; human figures; "medicine bags" or containers for the shamans' ritual paraphernalia (see photo 2); and a few snakes, mountain lions, and other beings. The human figures, especially the so-called patterned-body anthropomorphs, are individual portraits of the shamans as

Photo 1: A bighorn sheep petroglyph, Little Petroglyph Canyon. Engravings such as this constitute more than half of all of the petroglyphs in this region. They depict the special spirit helper of the Rain Shaman. (Scale: about 10 inches across.)

Photo 2: This complex petroglyph panel from Little Petroglyph Canyon includes depictions of shamans as patterned-body and solid-body anthropomorphs, various geometric patterns, the entoptic designs seen during trances, and skin "medicine bags" in which shamans kept their ritual paraphernalia. (Scale: central-right medicine bag is about 18 inches tall.)

Photo 3: The unique internal body designs in these patterned-body anthropomorphs from Little Petroglyph Canyon suggest individual "patterns of power" for different shamans. The spiral design in the central figure's face symbolizes concentrated supernatural power. The quail topknot feather headdress is specific to Rain Shamans. (Scale: central figure is about 3.5 feet tall.)

they saw themselves in the supernatural world of their trances. The faces on many of these engravings feature spirals or concentric circles, symbolizing the fact that, like the spiral, the shaman was a concentrator of supernatural power and that, in the supernatural realm, he could fly like the whirlwind (see site 8). Some of these shaman figures wear headdresses made from the curved topknot feathers of a quail (see photo 3). Quail topknot headdresses were the special ritual headgear of the Rain Shamans.

Chronometric dating in the Cosos suggests that some of the petroglyphs may have been made as long ago as 19,000 years, although most of the engravings appear to be 1,000 to 1,500 years old or less. Bow-and-arrow motifs postdate about A.D. 500, when these weapons first appeared in the region, while atlatl motifs (see site 24) apparently predate this benchmark. Although rare, occasional horse-and-rider petroglyphs are also present in the Cosos, indicating that some of the art was made during the historical period of recent centuries. The last documented shaman's vision quest in the Cosos occurred in the first few decades of the twentieth century.

The Coso petroglyphs fall within the boundaries of the China Lake Naval Weapons Center, at Ridgecrest, California. While the location of the rock art within an active military installation necessarily limits access to the sites, it also affords them a level of protection from vandalism that is unparalleled in the western United States. With only a few exceptions, the Coso sites are in pristine condition.

◢◣◢◣ **Visiting Site 1:** The only access to the Coso Range petroglyphs is through group tours conducted on weekends in the spring and fall, when the roads become passable and the weather conditions are not too extreme. The tours include Little Petroglyph Canyon, the largest and most spectacular of the Coso sites. Particularly notable are two large panels covered with engravings of elaborate shaman figures referred to as "patterned-body anthropomorphs"; both panels are on the right-hand side of the canyon as you walk downstream, south from the parking and covered picnic areas. A few small panels of painted motifs are also present in Little Petroglyph Canyon, about midway to its end, just below a now dry waterfall. And just beyond these pictographs is a very well-rendered atlatl, again on the right side as you head down the canyon. Tour guides can direct you to these and other notable panels within the canyon.

Tours of Little Petroglyph Canyon are conducted by the Maturango Museum (details are provided in the appendix). Reservations fill up months in advance, so it is wise to book a spot early. There is a fee for taking the tour, which lasts an entire day. This is one of the greatest rock art sites in the world, however, so it is truly worth the effort to visit.

SITE 2
Squaw Spring Petroglyphs
Johannesburg, CA

Spirits and Springs

A very small petroglyph site near the mining towns of Red Mountain and Johannesburg in northern San Bernardino County provides sharp contrast to the sprawling rock art of the Coso Range. Typical of an "average" petroglyph site in Numic territory, the Squaw Spring site most likely represents the activities of one or perhaps just a few shamans, recording their supernatural experiences at a vision quest locale in their home territory, perhaps near one of their seasonal villages.

Squaw Spring falls within historic Kawaiisu territory. The Kawaiisu were a Numic group ranging from the southern half of Panamint Valley and the Coso Range into the Tehachapi Mountains. This placed them on the boundary between the Great Basin/Numic cultures in the desert region and the south-central Californian cultures of the southern Sierra Nevada to the west. Perhaps because of their geographical position, the Kawaiisu were very much culturally transitional between the two regions. While in the Tehachapis, for example, they followed the subsistence practices of south-central California, which emphasized acorns as a dietary staple; but when they moved into the desert region during portions of the year, they necessarily followed lifeways more similar to their Great Basin neighbors. More to the point, though, they appear to have made south-central California–style pictographs when in the Tehachapi region and Numic-style petroglyphs when in the desert. This is evident by comparing the Squaw Spring petroglyphs with another Kawaiisu site, Tomo-Kahni (site 31), which is a painted cave near Tehachapi.

The Squaw Spring site also typifies the association between rock art and permanent sources of water—in this case, a permanent spring. Permanent springs were believed to be inhabited by supernatural spirits and were therefore places of great supernatural power as well as appropriate locales for visionary experiences (see site 9).

There are indications that, in some cases, shamans may have owned specific springs because of their intimate association with the spirits residing within the waters.

The limited number of petroglyphs at this site are entirely geometric in form, representing the entoptic patterns experienced by a shaman during the initial stage of a trance (see photo 4). We do not know what most of these different entoptic forms meant symbolically, and it is likely that their meaning was largely idiosyncratic to the individual shaman. Still, verbal descriptions of trances indicate that images of bright lights, stars, sunbursts, and so on were common as a shaman began to enter an altered state of consciousness. While it is tempting to impose our own astronomical interpretations on such motifs—a star-burst-like pattern, for example, interpreted as a star or as the sun—the verbal descriptions of shamans' altered states belie such facile cross-cultural identifications. This point is emphasized by our knowledge that Numic descriptions of the supernatural spirit Apo, the Sun, might include his appearance in a vision in the form of a human male or simply as a bright light. Entoptic rock art motifs are both the most common

Photo 4: This geometric petroglyph exemplifies one of the entoptic patterns a shaman might see during the first stage of his trance. Such designs served as "patterns of power" for shamans; the image might be painted on a shaman's ritual regalia in addition to being engraved on rocks. (Scale: about 6 inches across.)

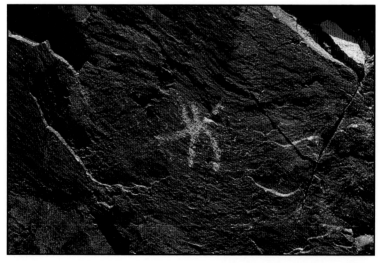

and the most enigmatic visions of the supernatural world left by the Native American artists.

In a strict sense, the age of the Squaw Spring petroglyphs remains unknown. The art lacks any well-developed revarnishing; however, when we combine this fact with the general temporal patterns of settlement in the region, it appears likely that they are less than 1,000 years old.

Visiting Site 2: Squaw Spring may be visited in conjunction with the Steam Well petroglyphs (site 3). You will need a four-wheel-drive vehicle for the short but bad road to the site, or you can hike about 2.5 miles from the main dirt road, which is suitable for two-wheel-drive cars.

To reach the site, take US Highway 395 north from Red Mountain toward Johannesburg and turn east (right) onto the paved Trona Road (see map 3). About 1.4 miles from the highway, turn right (southeast) again onto a dirt road on the south side of Trona Road. This is RM1444, a BLM road; it may not be identified as such at its intersection with Trona Road, but you will probably see a BLM plastic

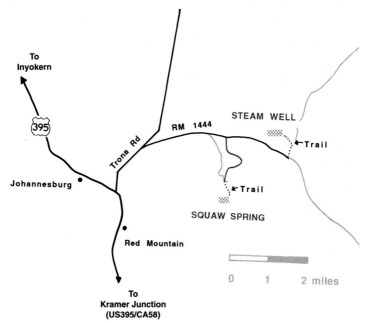

Map 3: Squaw Spring and Steam Well.

fence-post-type road sign at some point along the way. Squaw Spring is located about 1.25 miles south of RM1444 and is connected to it by a loop road turnoff. The first (western) arm of the loop occurs at 1.9 miles along RM1444 from Trona Road; the second arm of the loop hits RM1444 about 2.3 miles from the pavement. This second arm, which bypasses some small mine tailings and tunnels, is the better of the two turnoffs. It will take you to a fenced area marked "Squaw Spring Archaeological District." Go through the gate in this fence and walk downhill (south) on the abandoned road about 100 yards to a large concrete guzzler. The petroglyphs are located on a rock outcrop across a small arroyo, a few hundred yards southwest of the guzzler; a small BLM sign marks the outcrop.

The Squaw Spring site is only a few miles from Steam Well (site 3), so you can easily visit both sites in an afternoon. For road conditions, contact the BLM Ridgecrest Resource Area office (see appendix for the phone number). Because of the sometimes confusing pattern of dirt roads in this area, USGS topographical quadrangles (Klinker Mountain and Red Mountain 7.5-minute series) and/or the BLM "Red Mountain Desert Access Guide 7" map may be useful additions to the map provided here.

MAP 93
E·7

SITE 3
Steam Well Petroglyphs
Johannesburg, CA

Snakes and Bighorns

Steam Well is in the same vicinity as Squaw Spring (site 2) and is the larger and more elaborate of the two sites. Like Squaw Spring, Steam Well is located within Kawaiisu territory. Although its petroglyphs have not been dated using chronometric techniques, some appear to be about 2,000 years old while the majority are probably only half that age. Moreover, like Squaw Spring, this site is associated with a spring, further emphasizing the frequent connection between shamans' vision quest sites and permanent water sources.

The Steam Well petroglyphs consist primarily of entoptic patterns, but they include a few human figures and bighorn sheep. An unusual and particularly notable engraving at the site is a zigzag rattlesnake with horns—a combination of a rattlesnake and a bighorn (photo 5). Such conflations of different species, sometimes including humans (see site 4), into composite animals are common features of shamans' altered-states-derived art. They result from the third stage of a trance, when various figurative or representational hallucinations may combine with one another or with entoptic forms. What such composite beings may have meant in any specific sense is difficult to determine, again emphasizing the general fact that shamanistic experiences were often idiosyncratic in nature. The rattlesnake and the bighorn, however, were very important supernatural spirits for Numic groups like the Kawaiisu, and

Photo 5: A petroglyph depicting a schematized conflation of two animal species—the curving horns of a bighorn sheep combined with the zigzag body of a rattlesnake. Bighorn sheep and rattlesnakes were common spirit helpers for shamans in this region, and conflations of different animal species into a single fantastical being are common in early ethnographic accounts of shamans' visionary experiences. (Scale: about 12 inches tall.)

since they represent two of the most common identifiable motifs in the art, their combination into a single fantastical animal in a shaman's vision is not unexpected. Indeed, it signals the fact that the bighorn and rattlesnake were the most prominent spirit beings encountered by Numic shamans during their visions, and thus the most common animal species portrayed in their art.

◣◣◣ Visiting Site 3: An easy hike of about 1.2 miles round-trip is required. Steam Well may be visited in conjunction with the Squaw Spring petroglyphs (site 2).

The road to Steam Well is the same as that to Squaw Spring (see map 3). Take US Highway 395 north out of Red Mountain, then turn east (right) onto Trona Road. Turn right (southeast) again, onto BLM dirt road RM1444, about 1.4 miles from the highway. Follow RM1444 approximately 4.1 miles from Trona Road, or about 1.8 miles beyond the second turnoff to Squaw Spring, to a road branching to the left (north). This is the turnoff to Steam Well, which is located north of RM1444. Because the area north of RM1444 has been designated as wilderness, the road to the site has been closed at RM1444.

You must hike to the site, but it's an easy ramble along the abandoned dirt road; my seven-year-old daughter made the trek with minimal prodding and only a few stops to admire wildflowers, bugs, and other distractions along the way. Walk north on old road approximately 0.4 miles, where another dirt road intersects from the west (left). Follow this intersecting road west about 0.2 miles into a small arroyo in the hills. A large BLM sign marks the site area. Petroglyph panels are located on rock outcrops on both sides of the arroyo, with the panel containing the snake-bighorn conflation on the point of the outcrop to the south.

As with the Squaw Spring site, the pattern of dirt roads in this area can be confusing. Although not necessary for finding the site, the appropriate USGS topographical quadrangles (Klinker Mountain and Red Mountain 7.5-minute series) and/or the BLM "Red Mountain Desert Access Guide 7" map can be helpful. As of this writing, RM1444 to the turnoff/hiking path to Steam Well is adequate for a two-wheel-drive car; to check the current status of the road, call the BLM Ridgecrest Resource Area office (see appendix for the phone number).

SITE 4
Black Canyon Petroglyphs
Barstow, CA

The Shaman Transformed

The Black Mountain area north of Barstow is deservedly one of the better-known, even if heavily vandalized, rock art localities in the Mojave Desert. Falling somewhere between the massive concentration of petroglyphs in the Coso Range and the more typical small Mojave Desert sites found at many permanent springs, it consists of a series of small- to moderate-size concentrations of petroglyphs located along basalt canyon walls. Two sites on Black Mountain are open for visits: Black Canyon and Inscription Canyon (site 5). Black Canyon is the smaller of the two sites. You can easily visit it and Inscription Canyon in one day. In both cases, petroglyphs are found on darkly varnished basalt boulders on either side of sandy washes.

This area is thought to fall within the historical territory of the Kawaiisu, a Numic-speaking group of Native Americans who lived in the Tehachapis and central Mojave Desert. Related linguistically to the Paiute and Chemehuevi, the Kawaiisu appear to have maintained lifeways similar to these Numic neighbors in the Great Basin, to the east and north. In particular, their rock art is much like that of their northern and eastern neighbors.

At Black Canyon, the dirt access road runs through the middle of the wash containing the rock art panels. Most of the petroglyphs are concentrated on the lower boulders of the high scree slope to the east, as well as on the lower basalt outcrop to the west. Although dominated by the entoptic patterns commonly perceived during a shaman's trance, Black Canyon also contains human figures and bighorns, reminding us that the Coso Range was not the only location where shamans could obtain rainmaking power.

A north-facing panel on the east side of the canyon, just above the canyon floor, contains an unusual male bighorn (see photo 6). Instead of portraying the straight or slightly curved legs of a "normal" mountain sheep, this engraving shows the front legs bending

59

Photo 6: This petroglyph depicts a conflation of a bighorn sheep and a human (as indicated by the manner in which the legs and feet appear). A common spirit helper for Great Basin shamans, bighorns imparted to them the power to control rain. Once acquired, a spirit helper and a shaman became one and the same, and a shaman could transform himself into his helper when he entered the supernatural world. (Scale: about 12 inches long.)

forward from the elbows and the rear legs bending backward. Of course, this is anatomically impossible for a bighorn, but not for a person, suggesting that the art is more likely a conflation of a human and a bighorn. A smaller, partly defaced petroglyph is immediately behind this unusual bighorn; although the head is missing, it too appears to be a human-bighorn conflation.

Such conflations of different species are common in shamanistic arts. They are conceived when an individual in an altered state of consciousness begins to feel somatic hallucinations, such as the growth of hair or fur on the skin and the projection of antlers or horns out of the head, during the third stage of a trance. Such hallucinations are responsible for the widespread belief that shamans could transform, or "transmogrify," into their spirit helpers. In the southern Sierra Nevada, for example, shamans and grizzlies were so intimately associated that they were considered one and the same: a grizzly *was* a shaman transformed, and to be killed by a grizzly was to be murdered by an evil shaman. A similar association between bighorns and Rain Shamans prevailed in the Great Basin and the Mojave Desert. It is for this reason that the anthropologist Isabel Kelly stated in 1936 that "rain falls when a mountain sheep is killed. Because of this some mountain sheep dreamers [i.e., shamans who had visions of bighorns] thought they were rain doctors." That is, using a metaphor of "death" or "killing" for entering an altered state of consciousness, the Bighorn Shaman made rain when he entered a trance.

The theme of bodily transformation is also portrayed in a nearby panel, on the same side of the road but closer to the wash floor. This is a faint, north-facing panel that shows a human whose left side forms a zigzag fringe and whose left leg transforms into a long zigzag line. Zigzags, of course, were widely understood as schematized renderings of rattlesnakes, based on the zigzag pattern a sidewinder leaves as it slithers across the sand. In this case, this motif apparently portrays the transformation of a Rattlesnake Shaman into his supernatural alter ego; this power enabled the Snake Shaman not only to charm rattlesnakes, handle them with impunity, and cure their bites, but also to transform himself into a snake and travel through underground passages (see site 19).

No chronometric dating has been conducted in Black Canyon specifically, or in the Black Mountain region generally. Most of the motifs in Black Canyon appear to have a minimal amount of revarnishing, however, broadly suggesting that they may be less than about 2,000 years old and that most are probably less than 1,000 years old. Atlatl petroglyphs are present in the region, though, indicating that some of the art must predate bow-and-arrow technology of the past 1,500 years and thus could easily be a few thousand years old.

▟▚▞▚ **Visiting Site 4:** Your vehicle will need high ground clearance; four-wheel drive is recommended. Black Canyon may be visited in conjunction with the Inscription Canyon petroglyphs (site 5).

Access Black Canyon from California Highway 58 between Boron and Barstow (see map 4). Turn north on Hinkley Road, about 10 miles west of Barstow; it is paved most of the way. About 8.1 miles north of the highway you will encounter Brown Ranch Road (unpaved) on your left (west). Follow this road as it heads west and then turns to the north for a total of about 4 miles, where you will encounter Jackrabbit Road, an east-west dirt road that runs parallel to Black Mountain. Turn left (west) on this road and follow it approximately 2.4 miles while skirting a series of fenced alfalfa fields on your left (south). You will encounter another dirt road running to the north; turn right (north) on this road and go about 2.6 miles until it branches; stay to the right and you'll enter a narrow basalt canyon. The petroglyphs begin about 0.2 miles farther north, in the narrow portion of the canyon, and they stretch for about 200 yards on either side of the canyon.

The roads into and around Black Canyon are sandy and/or rocky. In good weather, a two-wheel-drive vehicle with high clearance, such as a full-size pickup truck or Volkswagen bus, can probably get into

Map 4: Black Canyon and Inscription Canyon.

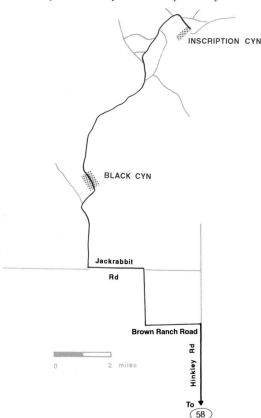

62

Black Canyon, but drivers must be very attentive and should carry a shovel for safety's sake. Obviously, a four-wheel drive would be more secure. Check road conditions through the BLM Barstow Resource Area office (see appendix for the phone number). Like many desert sites, there is a welter of roads in the Black Mountain region. Although you can find the site with the map provided here, the appropriate USGS topographical quadrangles (Lockhart, Bird Spring, Opal Mountain, and Water Valley 7.5-minute series) and/or the BLM "Red Mountain Desert Access Guide 7" map may serve as an aid in keeping you headed in the right direction.

SITE 5
Inscription Canyon Petroglyphs
Barstow, CA

MAPS 102-103
MAP93 G10

Supernatural and Cautionary Tales

Inscription Canyon may be the most spectacular Mojave Desert petroglyph site outside of the Coso Range. Located northeast of Black Canyon (site 4), it too falls in Kawaiisu territory. Though undated using chronometric techniques, the motifs at the site range from moderately revarnished to unvarnished, suggesting very generally that this location includes petroglyphs only a few hundred years old as well as some that may go back three to four thousand years or more.

The engravings are located at the mouth and along both sides of a low basalt-walled arroyo that empties from the south onto a dry lake bed. Motifs include bighorn sheep, human figures, and myriad geometric or entoptic forms. A very spectacular bighorn sheep engraving is located on the northwest side of the site, at the mouth of the arroyo, on a boulder near the ground (see photo 7). This mountain sheep is juxtaposed against an elaborate, rectangular herringbone pattern, which, judging from the degree of revarnishing, appears to be the same age as the bighorn and was probably made at the same time. As noted previously, one of the ways in which

Photo 7: A bighorn sheep petroglyph superimposed against a geometric design. During the third stage of a shaman's trance, iconic images such as a sheep might appear alone or sometimes superimposed on or against an entoptic design, such as this herringbone pattern. (Scale: sheep is about 2 feet long.)

images are commonly perceived during the third stage of a shaman's altered state involves the projection of a figurative form against or alongside a geometric pattern, as we see in this case.

"Patterned-body anthropomorphs"—human figures rendered with elaborate geometric designs for bodies—are also present at the site. Also resulting from the third stage of a shaman's trance, they represent the integration of entoptic forms with figurative images, creating complex stylized motifs. As in the Coso Range (site 1), there is no evidence of any repetition from one petroglyph to another of the internal body designs of these human figures; all of them appear unique. For this reason, they are most likely representations of individual shamans, engraved as the "power-form" the shamans assumed while in the supernatural world, with each different petroglyph serving as a kind of signature for a specific shaman. The tradition of shamans emphasizing their individual powers in such a manner appears to have become common after about 1,000 years ago, when shamans began to develop as the political and religious leaders of small bands of related families within the Mojave Desert and Great Basin regions.

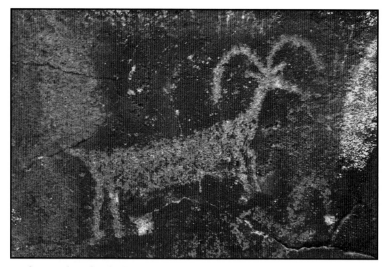

Photo 8: The red and yellow paint smeared around this bighorn sheep petroglyph shows the adverse effects of taking rubbings from the panel. Such practices degrade the integrity of the art as well as its scientific value, leading rock art site management agencies to enforce strict policies on visitation. (Scale: about 12 inches long.)

While there are many motifs worthy of examination at this site, one other panel warrants mention here, if only because it illustrates a cautionary tale. Located about halfway down the western side of the arroyo, this panel consists of a bighorn sheep with a dramatic band of red paint around it—the result of sloppy efforts to make a rubbing of this petroglyph (see photo 8). This panel is a textbook example of unintentional vandalism, and it presents a classic case of why petroglyph rubbings must be completely avoided: Not only has the paint ruined the aesthetic qualities of this rock art panel but it also contaminated the petroglyph so it is now undatable.

▲▲▲ **Visiting Site 5:** Your vehicle will need high ground clearance; four-wheel drive is recommended. Inscription Canyon may be visited in conjunction with the Black Canyon petroglyphs (site 4).

Inscription Canyon is most easily accessed from Black Canyon (see Visiting Site 4 and refer to map 4). From Black Canyon, continue north-northeast on the dirt road for roughly 7.8 miles (an approximate distance as the road winds through a sandy wash with many different tracks, all heading in the same direction) until you arrive at a small arroyo with low basalt walls, blocked by a large, low wooden fence made from telephone poles. The petroglyphs are located on both sides of the arroyo for a distance of about 100 yards.

As with the route to Black Canyon, cautious and experienced off-road drivers can make this trip with a high-clearance, two-wheel-drive vehicle, but you will be safer in a four-wheel drive. Check road conditions ahead of time through the BLM Barstow Resource Area office (see appendix for the phone number). Again, the map included here will get you to the rock art site, but you may find added security in the USGS topographical quadrangles (Lockhart, Bird Spring, Opal Mountain, and Water Valley 7.5-minute series) and/or the BLM "Red Mountain Desert Access Guide 7" map.

MAP 103
D 9-10

SITE 6
Surprise Tank Petroglyphs
Mojave Desert, CA

A Rattlesnake Shaman's Cave

Another important petroglyph site is at Surprise Tank in the Rodman Mountains. It falls within the territory of the Vanyume, a Takic-speaking group who lived in the foothills of the San Bernardino Mountains and north to the Mojave River. Surprise Tank contains a concentration of about 900 petroglyphs, making it one of the largest sites in the region. The engravings are located on both sides of a small basalt arroyo that drops unexpectedly into the desert floor; hence, presumably, the name "Surprise." Although there is no spring here, the basalt floor of the arroyo contains a series of natural tanks that trap water in pools during parts of the year, again illustrating the frequent association between rock art sites and water sources in the desert region.

A notable characteristic of the petroglyphs at this site is their relative degree of revarnishing: Some exhibit little or no revarnishing, and thus may be less than 1,000 years in age; but others, in great contrast, are heavily revarnished and appear almost as dark as the surrounding basalt rocks. Strictly speaking, it is impossible to use the degree of revarnishing as a definite indicator of age. Varnish accumulates around and within petroglyph engravings at differing rates in different regions of the desert; very localized conditions, such as exposure to water, affects the rate of varnish accumulation within a single site, even on an individual boulder. Still, a gross relative age may be suggested, based on the varnish accumulation rate in a particular region and on the revarnishing rate of other petroglyphs at the site. Cognizant of the fact that my hypothesis lacks the trial of chronometric testing, I estimate some Surprise Tank petroglyphs may date to the early Holocene—about 10,000 years before present—based on the relatively dark revarnishing of many petroglyphs at the site. Still, most are probably less than 1,000 years old. This signals the fact that places Native Americans recognized as sacred maintained that characteristic for millennia.

Like most petroglyph sites, entoptic forms seen during the initial stage of a shaman's trance are the most common engravings at Surprise Tank. This site also includes a significant number of human figures, bighorn sheep, hand prints and, especially, zigzag snake motifs. A particularly unusual series of panels is near the northwestern end of the arroyo (see photo 9). These consist of a number of zigzag and curvilinear snake motifs on panels that frame a small rock shelter. As mentioned elsewhere in this guide, Rattlesnake Shamans could handle snakes in ceremonies and even cure snake bites; supernatural snake spirits were their helpers, and zigzag lines represent a stylized technique for portraying rattlesnakes. This concentration of snake motifs around the small rock shelter most likely indicates the vision quest locale of one or more Rattlesnake Shamans.

Moreover, the rock shelter itself is significant: Caves often served as vision quest locales because shamans believed the supernatural world lay inside or beyond them; the shaman entered the supernatural when the rocks opened up for him. Caves served as portals to the sacred realm.

Across the arroyo from this Snake Shaman's cave, and a little farther down the wash, there is another interesting petroglyph: a

Photo 9: A small rock shelter within Surprise Tank surrounded by several small panels of petroglyphs. Rattlesnake motifs dominate these panels, the concentration of which suggests that this location was used by Rattlesnake Shamans for their vision quests. The rattlesnake spirit or specialized helper empowered shamans to cure snake bites. (Scale: shelter entrance is about 4 feet high.)

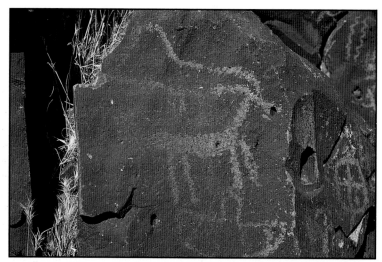

Photo 10: A bighorn sheep petroglyph apparently conceived when a shaman construed two long, meandering lines (typical entoptic patterns seen during a trance) as sheep horns. Such unnatural features in representational engravings supports the ethnographic indications that petroglyphs depict supernatural beings. (Scale: about 2 feet long.)

bighorn whose horns form sinuous lines longer than the animal's body (see photo 10). In the discussion of the neuropsychology of a shamans' altered state (see page 10), I noted that during the second stage of a trance the individual may interpret common entoptic forms as something meaningful or understandable within his own cultural context. Meandering lines, which can be sinuous or branching in form, are common entoptic patterns. We can infer, in this case, that a shaman perceived meandering lines simply as entoptic patterns created within his optical system due to his altered state of consciousness, and as a result of cultural conditioning and perhaps personal desires, he projected a meaningful form onto them — a bighorn, the spirit helper of the Rain Shaman.

▲▲▲ **Visiting Site 6:** You will need a high-clearance vehicle and/or the ability to hike about 3.6 miles round-trip.

Surprise Tank is located relatively close to Barstow, south of Interstate 40, but it is most easily approached from Lucerne Valley, to the south (see map 5). From California Highway 247 east of Lucerne Valley, head north on Camp Rock Road (paved). Follow Camp Rock Road about 4.1 miles, where you will reach a signed intersection with Granite Road entering from the east (your right) and Camp Rock Road angling toward the northeast; the northward extension of Camp Rock Road, from this intersection, turns into Harrod Road. Continue following Camp Rock Road, which shortly will become a dirt road, toward the Johnson Valley OHV (Off-Highway-Vehicle) Area. Approximately 14.1 miles from its intersection with Granite Road, Camp Rock Road intersects with Cinder Cone Road (BLM road OJ228, a dirt road), which angles in from the right (this will be the third dirt road that enters from your right; it is marked with a sign reading "Cinder Mine 7 Miles"). Turn onto Cinder Cone Road (OJ228) and follow it for a total of about 6.5 miles as it winds east and then north toward a large cinder cone. This should put you at an intersection with BLM road OJ233, which will be on your right. Turn right (east) on this road and follow it about 1.1 miles. You will notice some modern "petroglyphs" on a rock outcrop on your left soon after this road turns down a small arroyo. After you are about 1.1 miles from OJ228 you will intersect another dirt road heading north-northeast; follow this new road for approximately 0.7 miles, which will bring you to a large open flat

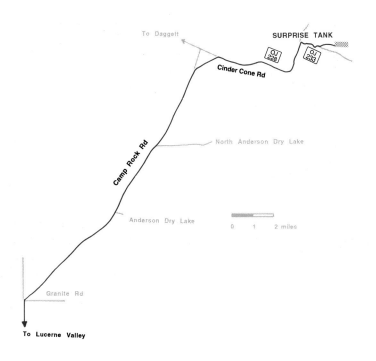

Map 5: Surprise Tank.

where you can park. Surprise Tank is directly ahead of you, but because it drops down onto the desert floor the site is not immediately visible from the parking lot. Walk over to the wash on your left (north) and follow it into the canyon. At one time an aluminum BLM fence ran across the wash, but as of this writing it had been cut down and carted away by vandals, presumably for its scrap value.

As you drive the last leg into the site, you may notice two fenced-off areas to your left, about 150 yards down a dirt track. These fences protect, in one case, an intaglio or geoglyph, and, in the other, what may be a Native American rock alignment. The intaglio is a curved design where the pavement of desert cobbles has been removed, exposing the lighter soil below. A stone "house-ring," or circle of cobbles used as the foundation for a brush hut, also lies within the first fenced enclosure. The second fenced area contains a possible rock alignment in the form of a boomerang; this may be a purposive arrangement of cobbles placed on top of the desert pavement.

Such earth figures are discussed in more detail later (see sites 20 and 22). Whether these Surprise Tank earth figures are aboriginal in origin is debatable. Although I doubt their authenticity, the presence of the stone house-ring supports an aboriginal origin. If they are genuine, they were likely used for some ceremonial purpose, although we have no ethnographic record of the manufacture and use of such art forms in this portion of the Mojave Desert.

Camp Rock Road and BLM road OJ228 are well maintained, but OJ233 is more rugged. The last half of OJ233 is very rocky; while it does not require four-wheel drive, your vehicle must have relatively high clearance. Alternatively, energetic visitors could easily walk the last mile or so to the site. Check road conditions with the BLM Barstow Resource Area office (see appendix for the phone number). The roads to the site are relatively straightforward and can be followed with the map included here; however, you may wish to take the BLM "Desert Access Guide 11" as an additional aid. Unfortunately, the relevant USGS topographical quadrangles—Silver Bell Mine and Camp Rock Mine—do not identify Surprise Tank or the roads leading to it, so they are of little use here.

SITE 7
Fish Slough Petroglyphs
Bishop, CA

MAP 77
A-C 9

Sacred and Profane Space

Four petroglyph sites (7, 8, 9, and 10) lie in the upper Owens Valley, near Bishop. All are open to the public and easily accessible, and together they form a "petroglyph loop tour" that can be completed in one day. All of the sites fall within the territory of the Owens Valley Paiute, a Numic-speaking group of Native Americans closely related to the Coso Shoshone, who lived farther to the south. The four sites in this loop tour provide good examples of the range of variation in Numic rock art. They contrast, in some cases dramatically, with the Coso petroglyphs (site 1), which are also Numic in origin.

The first stop on the loop tour is Fish Slough, a small site consisting of a few dozen petroglyphs. As with the other three sites, these have been engraved in volcanic tuff, a relatively soft stone that can be pecked (or engraved) easily but also erodes quickly, geologically speaking, and often does not develop or retain rock varnish for any lengthy period. Due to the nature of the rocks, it is unlikely that any of the petroglyphs at Fish Slough, as well as the other three sites on the loop, are more than a few thousand years old; most are probably less than a thousand years in age.

All of the motifs at Fish Slough are entoptic patterns, the geometric designs perceived during the first stage of a shaman's altered state of consciousness (see photo 11). Bisected circles (with vertical lines drawn through them) are particularly common at this site. Some archaeologists have suggested that bisected circles represent schematized atlatl motifs (see site 24); others say they may portray vulva-form motifs (see site 10). In both cases, the bisecting line was always drawn vertically, suggesting an intentional and therefore standard meaning. Some of the motifs indeed seem to suggest atlatls, especially when the vertical bisecting line is quite long and the circle appears at its lower end, with the circle representing finger loops of the throwing board while the long line represents the "arm" of the weapon. When the bisecting vertical line is the same length as the diameter of the circle, in contrast, the vulva-form identification may be correct, although we cannot be certain.

While the ethnographic record provides us with a good understanding of the origin of the Numic petroglyphs, it does not necessarily yield interpretations of all of the motifs. We can identify many, if not all, of the geometric motifs as variations of entoptic designs. We know that during the second stage of a trance shamans sometimes construed entoptic patterns as culturally meaningful figurative visions, such as curved parallel lines meaning bighorn sheep horns; and we can understand that specific cultural meanings may have been assigned to particular entoptic motifs, like zigzags and diamond chains to represent rattlesnakes. But we do not know what all of the entoptic patterns mean and, since each shaman's trance

Photo 11: Typical of most rock art sites in the Great Basin, the entoptic designs on the horizontal panel in the foreground dominate the motifs at Fish Slough. Shamans saw such designs during the first stage of a visionary trance. The commonness of these motifs alone at various sites belies the popular belief that rock art was produced for "hunting magic."

was to some degree idiosyncratic, it is unlikely that we will ever decipher all of the geometric forms beyond recognizing them as patterns of power.

Grinding slicks—flat, smoothly polished surfaces on horizontal boulder surfaces, used for grinding seeds and nuts—are also common at Fish Slough, indicating that some rock art sites served as multipurpose locales: Though used at certain times for shamans' vision quests, they might also be used at other times for more mundane activities, such as grinding seeds. Thus, shamans' vision quests were not necessarily conducted at remote locations but might occur, instead, within a seasonally occupied village. The Numic did not necessarily make a strong spatial distinction between the location of sacred activities and those where mundane (secular or "profane") acts would be conducted. Curing rituals, for example, would be undertaken within the sick individual's hut; circle dances were performed within the village; and even the dead would be buried under the floor of their hut, which was then burned. Rock art sites, of course, were associated with places on the landscape imbued with supernatural power, but the important point here is that, often, so were villages and other site areas. Unlike in our culture, which sets its sacred places apart from the mundane areas of daily living, for the Numic the supernatural and spiritual were everywhere around them in their day-to-day activities.

▰▰▰ **Visiting Site 7:** To take advantage of the "petroglyph loop" tour, visit Fish Slough in conjunction with Chidago (site 8), Red Canyon (site 9), and Chalfant (site 10).

Due to incidents of vandalism, the BLM has decided, with support of local Native Americans, to control access to the Bishop sites, including Fish Slough, in part by not widely disseminating directions to them. They have asked that directions to the sites not be printed in this guide; however, you may obtain directions and a map to these sites from the BLM office in Bishop, which is located in the Cottonwood Plaza, 785 North Main Street, Suite E, Bishop, CA, 93514. North Main Street is US Highway 395 through town and is directly en route to the sites. The BLM hopes that by requiring individuals to appear at its office to obtain maps it can prevent vandals from gaining access to the sites.

Follow the BLM map mileage figures closely because, although the sites are not hard to find, some the roads and turnoffs are not marked. It is also a good idea to follow the BLM mapped order for the sites, which leads you logically from small to increasingly larger and more complex petroglyph sites. Fish Slough is located about 100 yards up a small unmarked turnoff, which will be on your left; look for a large cattle corral to the right of the road just before the turnoff to the site. The parking area is marked by low wooden barriers, with the petroglyphs on low tuff boulders and outcrops just beyond the parking lot.

SITE 8
Chidago Petroglyphs
Bishop, CA

The Power of the Whirlwind

The second Owens Valley Paiute site on our Bishop petroglyph loop tour is Chidago. Slightly larger than Fish Slough, the Chidago site contains roughly a hundred motifs within a relatively restricted pile of tuff boulders. Like most rock art sites, the large majority of the motifs at Chidago are entoptics, the geometric patterns perceived during the initial stage of a shaman's trance, although a few representational forms are also present, including a possible lizard. As at Fish Slough, bisected circles are present, along with grids, zigzags (rattlesnakes), and a few carefully rendered spirals and concentric circles (see photo 12).

The spirals and concentric-circle motifs are particularly notable since they are one of the few "geometric" motifs to which we can assign a specific cultural meaning. As noted above (see site 1), these are the whirlwind designs that represent concentrators of supernatural power. Throughout far-western North America, whirlwinds were believed to contain ghosts, a particular kind of supernatural spirit. Whirlwinds were also closely associated with the shaman's power to fly, which was a metaphor for entering an altered state of consciousness, resulting from the somatic hallucination of weightlessness that sometimes accompanies trances. Indeed, among the nearby Numic-speaking Western Mono in the southern Sierra Nevada, the terms for "spirit helper" and "flying power" could be used interchangeably, while a recorded historical account of an Owens Valley Paiute shaman's vision begins when he is sucked up into the sky by a whirlwind. The Numic Circle Dance performed at most social gatherings inscribed a circle or series of concentric circles on the landscape. It, too, was intended to concentrate supernatural power for the benefit of the community. And, of course, one of the most common forms of human heads on Numic shaman petroglyphs is the stylization of the face as concentric circles or a

Photo 12: This complex panel of petroglyphs illustrates the numerical importance of entoptic motifs in the art at most sites. Note the concentric circle in the upper-middle portion of this panel; such designs represent the whirlwind, which shamans believed contained spirits and could carry them into the supernatural world, thus concentrating supernatural power. Since shamans also concentrated power, the substitution of a spiral or concentric circle for their face is common on many of the human-figure motifs. (Scale: circle diameter is about 8 inches.)

spiral, because, like the whirlwind, the shaman too is a concentrator of supernatural power.

The concentric circles and spirals at Chidago, like the other entoptic patterns engraved at this site, are symbols of supernatural power. But in a more specific sense they symbolize the fact that the shaman who conducted his vision quest at this location was an individual who concentrated power, which he used for curing and other ritual activities.

▰▰▰ **Visiting Site 8:** Following the "petroglyph loop tour," visit Chidago in conjunction with Fish Slough (site 7), Red Canyon (site 9), and Chalfant (site 10).

For directions and a map to the Chidago petroglyphs, you must personally visit the BLM office in Bishop (see Visiting Site 7). The site is in a chain-link fenced area immediately on the right side of the road and, following the BLM instructions, is easy to spot.

SITE 9
Red Canyon Petroglyphs
Bishop, CA

The Path of the Water Baby

The third stop on the Bishop loop tour of Owens Valley Paiute art is at Red Canyon, a fine example of a medium-size petroglyph site. It has two concentrations of petroglyphs: a small group around a parking area on the western side of the site; and a larger group just beyond another parking area about 200 yards east of the first car park (see photo 13). Both areas are worth visiting.

In addition to the entoptic patterns common at all rock art sites, Red Canyon contains figurative motifs scattered in and among the

Photo 13: A panel of entoptic patterns, zigzags, and other motifs in the eastern portion of Red Canyon, with the White Mountains in the background.

complex panels. There are at least three human figures: two stick-figure "digitates"—displaying prominent, widely splayed fingers and/or toes; and one very wide and deeply pecked solid-bodied human, engraved on an almost-horizontal panel near the eastern car park. The site also includes images of a few bighorn sheep and a wide variety of zigzag snakes. Most notable, however, are the panels covered with hand and especially footprints (see photo 14). The handprints are located on the vertical face on a boulder in the western concentration of petroglyphs, while the footprints are on high horizontal panels in both site areas. You may have to climb up on the rocks to see them, so be careful not to step on or damage any motifs. A set of two footprint panels in the western site area contains human footprints, placed as if walking across the rock, and a distinctive set of three bear paw prints heading in the same direction.

Footprints (or tracks) and handprints are common in rock art sites worldwide (see site 26). Tracks played an important role in

Photo 14: A panel of engraved footprints. Prints and tracks are common at hunter-gatherer sites worldwide, attesting to their general communicative importance for hunting peoples. In the Great Basin, small human footprints were attributed to the Water Baby, a diminutive human spirit who lived in springs and pools and served as a particularly potent spirit helper for shamans. The sight of Water Baby's tracks was thought to signify a supernatural experience. (Scale: most footprints are about 6 inches long.)

hunting-oriented societies, forming a logical and readily recognizable symbol for any particular animal species. Among the Numic, however, humanlike tracks were particularly associated in supernatural visions with an appearance of "Rock" or "Water Baby," an unusually potent shaman's spirit helper. The association between a shaman and this spirit helper was so great, in fact, that many Numic informants stated that "water baby made the petroglyphs" when queried by an anthropologist about their origin, in this case alluding to the inseparability of the actions of a shaman and his helper.

Several informants described Water Baby—a more accurate translation might be "Water Dwarf"—as a short, long-haired male human wearing traditional Native American clothing. They believed he lived in springs and rivers and, like all supernatural spirits, was fond of native tobacco, a strong hallucinogen. The sighting of Water Baby was believed to result in death—a metaphor, in fact, for entering or being in an altered state of consciousness. Most commonly the informants said Water Baby's footprints could be seen around the springs and water holes where he resided.

One of the interesting characteristics of the Water Baby is his diminutive size, not unlike the "little people" common in myths and legends worldwide. In the case of the Numic, and perhaps many other cultures, the size of the Water Baby appears to result from a common reaction to an altered state of consciousness, referred to as a "Lilliputian hallucination." Apparently due to changes in the optical system during altered states, and sometimes from the frequent ingestion of hallucinogens, a person may experience visions that make things seem smaller than they really are; hence, the belief in a small-statured spirit. Many of the small anthropomorphic figures engraved or painted at numerous rock art sites probably share this same origin.

▰▰▰ **Visiting Site 9:** Following the "petroglyph loop tour," visit Red Canyon in conjunction with Fish Slough (site 7), Chidago (site 8), and Chalfant (site 10). All sites on this tour require obtaining directions and a map at the Bishop BLM office (see Visiting Site 7 for details). The Red Canyon site is on the north side of an H-shaped intersection of dirt roads, on a series of low boulders and volcanic tuff cliffs.

MAP 77
A-C 9

Visiting the Sites

SITE 10
Chalfant Petroglyphs
Bishop, CA

The Power of the Vulva

The fourth and final Owens Valley Paiute site on the Bishop petroglyph loop tour is the largest and most spectacular, containing over 400 petroglyphs spread along 600 yards of cliff face. Like the other sites in this region, the soft chalklike texture of the tuff cliff erodes easily. It is rapidly sloughing off and collapsing in some areas, while being scoured smooth by the wind in others. The implications of dating the art in this setting are straightforward: It is unlikely that any of the petroglyphs here are more than a few thousand years old, and most are probably 1,000 years or less in age.

The Chalfant site stands out partly because of the commonness of a single motif: bisected circles, which are numerically the largest category of petroglyphs at the site. As noted in our discussion of site 7, some bisected circles may be identified as "vulva-form" petroglyphs; those at the Chalfant site fall within this group (see photo 15). This is particularly apparent in some of the motifs, which were made by incorporating natural declivities and other features of the cliff face to create what are almost low relief, three-dimensional vulva sculptures.

Several decades ago archaeologists suggested the vulva-form motifs in this region may have resulted from some form of girls' puberty ceremony, as did other such art in southwestern California. Although a logical suggestion at the time, this hypothesis was plausible only when we incorrectly believed that no ethnography existed to enlighten our understanding of the art, and when we thought the petroglyphs were very ancient. We now know the ethnographic record contains considerable information about the origin and meaning of the petroglyphs. And we know much of the art is recent rather than ancient in age. Perhaps most important, we have found no ethnographic evidence linking the creation of rock art with any kind of girls' puberty ceremony among the Numic-speaking peoples of the Great Basin. We have found in the ethno-

Photo 15: Petroglyphs at the Chalfant site are engraved into soft volcanic tuff. Note the lower bisected-circle motif, one of the many vulva-form engravings common at this site. These symbolize a sexual association with the shamans' altered-state experiences, equating their entry into the supernatural with sexual intercourse; the sites themselves symbolized vaginas. (Scale: vulva form is about 4 inches.)

graphic record, however, considerable information about the origin of the art and its important connection to far-western Native American sexual symbolism. Although complex, understanding the sexual symbolism of shamanism is critical to interpreting rock art and is required to grasp the meaning of the vulva-form motifs at the Chalfant site.

Fundamental to the sexual symbolism of the shaman was the metaphysical belief that the supernatural world was intimately tied to the fecundity of the natural world, and through the shaman's manipulations of supernatural power he could influence conditions in the mundane. Because of the obvious association between sex and fecundity—and the perceived relationship between fecundity and supernatural power—it was believed that supernatural power was manifest in sexual potency. Moreover, there was a direct physiological link between sexuality and entering the supernatural world through an altered state of consciousness; some hallucinogens used by Native Americans, such as jimsonweed, are also aphrodisiacs. Shamans used sexual intercourse as a metaphor to describe entering the spirit world (see site 32). They expressed it symbolically by using an exaggeratedly phallus-shaped pestle in the ritual preparation of hallucinogens, and also by believing that "wet dreams" among nonshamans in their normal sleep were the result of dangerous sexual intercourse with spirits.

In both belief and practice, shamans were men of great sexual appetites whose predatory nature was widely recorded in the ethnographic literature. Young girls were warned to avoid the sexually rapacious shamans whenever possible because they might concede to the shamans' advances to avoid supernatural retribution directed toward them or their family. Shamans throughout the Far West were both feared and respected, reflecting a fundamental Native American belief about supernatural power: It was inherently amoral and ambivalent, and thus could be used for good or bad. This made power dangerous to those lacking the training and ability to use it, and since the shaman was a man of power, he was necessarily a dangerous individual. Following the same logic, any inherently dangerous creature—such as rattlesnakes, grizzlies, and spiders—was also supernaturally powerful and was therefore intimately associated with shamans. The symbols of shamans were likewise potentially dangerous because of their material spirituality connecting them to the sacred (see site 30).

Perhaps it is not surprising, then, that the vulva itself was considered unusually perilous. For example, a Northern Paiute account indicates that the worst form of sorcery a man could endure was a twitching vulva during intercourse: Female orgasm was thought to represent uncontrolled sexual, and therefore supernatural, power. Similarly, even the sight of a vulva posed a particularly dangerous circumstance. Another Northern Paiute incident tells of a public health nurse in the early 1900s who sought to promote hygiene by conducting a healthy baby contest on a reservation. Unceremoniously, she undressed the babies for a health inspection before the assembled reservation population. The hall quickly emptied as soon as the nurse disrobed a female baby—an unacceptable breach of etiquette because of the potency, and therefore danger to the people in the audience, of an exposed vulva.

The vulva, a potent and dangerous object, was an appropriate shamanistic symbol for supernatural power, perhaps pertaining to sorcery. That it is obviously female in nature might suggest that the vulva was associated with female rather than male shamans. While it is possible that vulva forms were the marks of female shamans, the evidence suggests the art, most likely, was made by males. I discuss the question of female shamans elsewhere (see site 21), but suffice it here to note that they were rare and do not appear to have been major contributors to the existing corpus of rock art. More to the point, however, is the fact that when females were known to have made rock art—specifically in the girls' puberty ceremonies of southwestern California (see sites 17, 35, 36, and 37)—the spirit helper they received, and the pattern they painted on the rock, was the rattlesnake. That spirit was widely associated with females throughout the Far West, with the corresponding zigzag and diamond-chain motifs universally taken as "feminine designs."

The importance of the female emphasis on rattlesnake designs in rock art, as well as in other art forms such as basketry, becomes apparent when the symbolism of the snake is considered (see site 21). At the most obvious physical level, the snake is a phallic, masculine symbol; so the most phallic species on the landscape was used by females. The significance of this symbol as a feminine design represents a symbolic inversion: Its importance is emphasized precisely by its physical contrast with what would normally be considered feminine. In a similar manner, rock art sites were themselves *female* places on the landscape, with caves and rock shelters being analogous to vulvas and vaginas (see site 32), yet they were largely used by male shamans. In both cases, these inversions allude to a basic belief about the supernatural world: that it was the exact opposite of the mundane. A male shaman who wanted to enter the most masculine portion of the supernatural sought its inverted counterpart in the natural world, a feminine place. And females who wanted a spirit helper to strengthen their femininity in the natural world sought the most masculine spirit of the supernatural, the rattlesnake, as their protector.

There are obvious ideological implications for Native American gender roles and statuses that underlie this sexual symbolism, but they need not concern us here. Given the overall structure of far-western Native American symbolism, beliefs, and metaphysics, it is most likely that the vulva-form petroglyphs at the Chalfant site were made by male shamans following their vision quests. On one level, the art symbolizes the shamans' entry into the supernatural by allusion to sexual intercourse as a metaphor for an altered state; on another level, it tells us the sites were feminine places generally used by male shamans. The art may pertain specifically to sorcery, due to the association between the vulva and evil supernatural power.

In addition to the vulva-form motifs at this site, there are a number of other designs, the most spectacular of which are a series of large circular motifs internally decorated with geometric patterns. Most likely, these are some form of complex entoptic pattern, although any specific meaning they may have had is now lost.

Visiting Site 10: Chalfant, the fourth and final site in the Bishop "petroglyph loop tour," may be visited in conjunction with Fish Slough (site 7), Chidago (site 8), and Red Canyon (site 9). All sites on this tour require visitors to obtain directions and a map at the Bishop BLM office (see Visiting Site 7 for details).

Follow the BLM map carefully. After the final turn in the BLM directions, stay on the primary dirt road for roughly a quarter of a mile, which will bring you to a parking area marked by low wooden barriers roughly in the middle of the site. The petroglyphs spread out on the tuff cliffs immediately across a small creek, which will be in front of you. Some of the most spectacular panels are located in a fenced-in area toward the northeastern end of the site.

MAP '79
H-8

SITE 11
Titus Canyon Petroglyphs
Death Valley National Park, CA

Making Do with Less

Death Valley, in name as well as in recent history, has become a symbol for the most inhospitable environment in the country. Aside from a small resident population of National Park Service personnel and others involved in the tourist trade, the valley is essentially

uninhabited and seemingly uninhabitable, at least from the perspective of what makes logical economic sense. Yet this apparently
desolate region sustained a stable, if small, Numic-speaking population of Shoshone and Southern Paiute, despite the fact that they
did not employ many of the technological advantages, such as irri-
gation and agriculture, that we would consider necessary to live in
this environment. This, of course, is because the Numic speakers
relied on their sophisticated, intimate knowledge of the region's
natural environment and resources rather than tools and machines.
Although many early travelers, including some anthropologists,
derided Numic culture as "primitive" and "simple," they judged
the natives on a purely materialistic scale. Others, who value environmental knowledge and mental and cognitive skills over material items, might conclude that the Numic inhabitants of the Death
Valley region were among the world's most sophisticated and successful people, precisely because they were able to succeed in such
a seemingly harsh place, almost entirely on their wits.

A small petroglyph site lies in the lower reaches of Titus Canyon, below the Grapevine Mountains that form the northeastern
side of Death Valley. Like many small Numic sites, the petroglyphs
are adjacent to a small spring, reflecting the widespread belief that
water sources provide entries into the supernatural and are inhabited by spirits (see site 2); reeds, and even a few palms, grow in this
area. The few motifs at the site depict a wide range of petroglyph
types: a stick-figure human, a bighorn, zigzag snakes, and, of course,
a variety of entoptics (see photo 16). The presence of a bighorn
petroglyph, representing the spirit helper of a Rain Shaman, points
to the fact that weather control, as well as other kinds of supernatural powers, could be obtained at many locales, even while certain areas, such as the Cosos (site 1), were well known for specific
shamanic specialties. None of the motifs at Titus Canyon look particularly old—that is, there is no significant amount of revarnishing—
suggesting that they are likely less than 1,000 years old.

The Titus Canyon petroglyphs are located immediately east of
Titus Spring, on a large boulder alongside the road. A prominent
sign marks the site, with the primary panel and majority of the
motifs on the west side of the boulder to the left of the sign. Incidentally, don't believe the Park Service sign that claims no one

Photo 16: A complex panel of petroglyphs at the Titus Canyon site. As is typical of many smaller petroglyph localities in the Great Basin, this site is immediately adjacent to a spring. Shamans conducted vision quests at such locations because of their belief that springs and other permanent water sources were inhabited by supernatural spirits, but they used them only during times when the spring area was otherwise uninhabited by villagers. (Scale: stick-figure human is about 12 inches tall.)

knows why the art was made or what it represents and proclaims that one person's guess is as good as another's. By now, your interpretation of this art should be much better than most people's.

Although the twenty-seven-mile drive through Titus Canyon is relatively long, about two hours, it is one of the most spectacular in the state. Visiting the rock art site is simply an added inducement. Named after Morris Titus, a prospector who died in 1905 along with two other men while trying to find water, the lower stretches of the canyon, below the site, is a sinuous passage through high vertical rock cliffs that are particularly picturesque. Indeed, early brochures for investment in Leadfield, now a ghost town along the road, touted Titus Canyon as California's Grand Canyon. Like the mining stocks the brochure intended to promote, this claim too was overblown; but the drive is dramatic, nonetheless.

/\/\/\ Visiting Site 11: You will need a four-wheel-drive vehicle to access this site.

A one-way dirt road runs through Titus Canyon (see map 6). It begins on Nevada Highway 374, approximately 4 miles west of Beatty. The petroglyphs and spring are about 18.4 miles from the highway; 8.6 miles beyond the site, the dirt road intersects with the paved road that runs through Death Valley proper. The dirt road is generally quite good and well maintained; although sandy in some spots and worn to washboard in others, you should be able to negotiate it in two-wheel drive. In the lowest stretch of the canyon, however, you

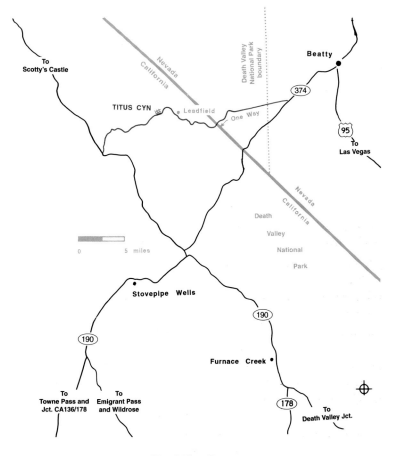

Map 6: Titus Canyon.

will encounter a very narrow, sinuous rock chute. The canyon walls tower hundreds of feet overhead and there's no place to turn around, let alone drive to higher ground. Avoid this road in rainy weather, or even if thundershowers look threatening, due to the danger of being trapped in a flash flood.

MAP 110
C-3

SITE 12
Pleasant Valley Petroglyphs
Joshua Tree National Park, CA

Geology and Rock Art

Two small petroglyph sites lie within Joshua Tree National Park, which places them in the Desert Serrano territory that ranged from the San Bernardino Mountains into the Mojave Desert as far north as the Mojave River. In some respects, the Desert Serrano, a Takic-speaking people, were transitional between the Numic speakers of the Great Basin, to the north, and other southwestern Californian Takic-speaking groups, such as the Cahuilla in inland Riverside County, to the south. As with most of their cultural characteristics, the rock art made by the Serrano nonetheless generally corresponds to the patterns in rock art production of southwestern California: It was created either by shamans or by puberty initiates. Both of the sites in Joshua Tree National Park, however, appear to be the work of shamans rather than puberty initiates.

The Pleasant Valley petroglyph site contains about a dozen petroglyphs dispersed across a boulder-strewn hill (see photo 17). As is typical of most sites, entoptic designs dominate, although there are two possible representational engravings as well: one is a schematically drawn human; the other is a zoomorphic figure, perhaps a lizard—the messenger to the supernatural world.

Two aspects of the location of this site are worth noting. First, the site is situated in the only area of darkly varnished rock (in this case, quartz monzonite) in the immediate region; everywhere else

Photo 17: The Pleasant Valley site is on a boulder-strewn hillside. About a dozen petroglyphs, probably made by Desert Serrano shamans, are dispersed on boulders such as the one pictured here. Like many small desert sites that archaeologists often overlook in favor of the few massive concentrations of art, this site is associated with a nearby village and water source.

the rock outcrops are lightly colored with little or no varnishing, thus providing little advantage to the Native American artist. Second, and perhaps more important, this area of darkly varnished rocks is close to the Serrano village of Squaw Tank, about 1,000 yards to the west. If the petroglyphs were made when the village was occupied, after about A.D. 1000, we can then infer that the art is less than a thousand years old. Since the petroglyphs contain little or no revarnishing, this is likely.

Visiting Site 12: You will need a four-wheel-drive vehicle. Pleasant Valley may be visited in conjunction with the Barker Dam petroglyphs (site 13). You must pay a fee to enter Joshua Tree National Park.

The Pleasant Valley petroglyphs are located on the Joshua Tree National Park Geology Tour Road (see map 7). From the paved Queen Valley Road, take the dirt Geology Tour Road south. The ninth stop on this tour, about 5.2 miles from the pavement, is the Serrano village of Squaw Tank. The tenth stop, approximately 0.4 miles farther, is the petroglyph site. The primary concentration of

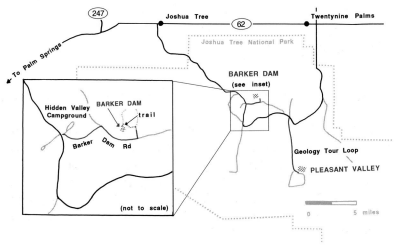

Map 7: Pleasant Valley and Barker Dam.

petroglyphs is located on a series of medium-size boulders about 150 feet north of the road. A few small geometric motifs are scattered sporadically to the east within this boulder field. A "Geology and Man" pamphlet, providing a self-guided tour of the geology loop, is available for a small fee at the Visitors' Centers. For more information, contact Joshua Tree National Park, 74485 Palm Vista Drive, Twentynine Palms, CA 92277; the phone number is listed in the appendix.

MAP 110
B-2

SITE 13
Barker Dam Petroglyphs
Joshua Tree National Park, CA

Art and Conservation

Another example of Desert Serrano region petroglyphs is located in the Barker Dam area of Joshua Tree National Park. Although the site contains only about two dozen motifs, it provides a good example of the variety—entoptic and representational art—

that sometimes occurs at vision quest locales. Unfortunately, the petroglyphs at this site were vandalized some decades ago by a movie company that painted them in garish colors so they would show up better on film.

Although it is difficult to ignore the thick bright green and red paint on the petroglyphs, this interesting site is worth the effort to see. The motifs include circle-headed, stick-figure humans, a human with a bow, two bighorn sheep, a possible tortoise, various snakes, and a number of entoptic designs such as double diamonds and circles (see photo 18). The bighorn may have been a special spirit helper for the Rain Shaman among the Serrano, as it was among the Numic-speaking peoples. Supernatural rattlesnakes, however, were important to all far-western North American groups, either as a kind of specialized spirit helper and/or as a dangerous

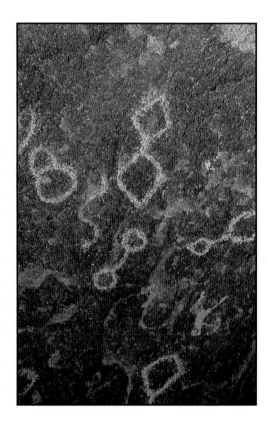

Photo 18: The double-diamond chain and other entoptic designs dominate petroglyphs at the Barker Dam site. Native Americans throughout far-western North America used the diamond chain to represent the scale pattern on diamondback rattlers and the zigzag to represent the trail sidewinders left in the sand. Such designs are common at rock art sites because rattlesnakes were important inhabitants of the supernatural world, serving both as important spirit helpers and, in some cases, as guardians of the supernatural. (Scale: double-diamond chain is about 12 inches.)

guardian of the supernatural. As noted many times in this guide, graphic conventions for depicting rattlesnakes included zigzags and diamond-shaped chains. Two large end-to-end diamond motifs at the site, combined with zigzags and a seemingly very representational snake motif, attest to the importance of this spirit among the Serrano shaman(s) who made art at this site.

The petroglyphs at Barker Dam are in a small cavelike alcove within a large boulder. The paint has eroded from the lower petroglyphs at the site, which sit on an almost horizontal floor, suggesting that the paint on the other motifs may be removable. This raises important questions about whether any previous effort has been made to remove the paint and how to go about it while safeguarding the art. As far as I have been able to determine, the National Park Service has made no attempt to remove the paint. Presumably, there are two primary reasons for this. First, restoration is a delicate and complicated task that requires the skill of a trained rock art conservator. The conservator must first identify the paint's chemical constituents to determine how it has bonded to the rock face; the next step would be to develop a plan for removal that ensures that the restoration process does not harm the engravings. For example, pigments may penetrate the rock face and actually become part of it; removing them might require removing part of the engraving, which in many cases would not be justifiable. The subtext of all of this, of course, is that restoring the site would be expensive, which brings us to the second reason I presume no restoration has been attempted: Restoring the site has not been a priority for Joshua Tree National Park. This might change, of course, if it becomes apparent to the Park Service that the conservation of this rock art site is important to the visiting public. So I encourage you to make your wishes about the importance of conserving rock art sites known to the authorities.

Visiting Site 13: A short but easy hike is required. Barker Dam may be visited in conjunction with the Pleasant Valley petroglyphs (site 12). You must pay a fee to enter Joshua Tree National Park.

You can reach the Barker Dam petroglyphs from the Hidden Valley campground within Joshua Tree National Park (see map 7).

Immediately after turning off the main road to the campground, turn right onto Barker Dam Road, the first dirt road in the camping area. Follow the signs toward Barker Dam. The turnoff to the dam will be on your left, about 1.5 miles from the main road. Walk north from the end of the parking lot, which is 0.2 miles north of Barker Dam Road. You will see a signed fork in the trail about twenty yards north of the parking lot. Take the fork to your left (west). The rock art site, which is well signed, is about 0.3 miles along this easy trail. For information, contact Joshua Tree National Park, 74485 Palm Vista Drive, Twenty-nine Palms, CA 92277; the phone number is listed in the appendix.

MAP 109
D-E 10
Contact Tribal Office

Colorado Desert Region

SITE 14
Andreas Canyon Pictographs
Palm Springs, CA

The Power of the Rock

A small but impressive site containing pictographs, petroglyphs, and cupules is in Indian Canyons on the Agua Caliente Reservation, on the southwestern side of Palm Springs. The Agua Caliente are a band of the Cahuilla, a Takic-speaking group whose traditional territory ranged across the San Jacinto Mountains from the Coachella Valley and into what is now western Riverside County. Typical of southwestern Californian groups, the Cahuilla made rock art in two different contexts: by shamans following their visionary experiences, and by puberty initiates at the conclusion of their rites.

Because the Cahuilla straddled the Colorado Desert and inland coastal regions of southern California, their rock art is stylistically transitional between the desert region, which emphasizes petroglyphs, and coastal southern California, where pictographs are more common. Andreas Canyon provides a good example of this mix, displaying both kinds of art. Works by puberty initiates, however, are not represented here; ethnographic evidence indicates this site was made by a shaman, or group of shamans, probably in the last few hundred years.

The site consists of a low rock shelter formed by a jumble of large slabs of rock (see photo 19). A series of faint pictographs — including a large set of concentric circles, a double-diamond motif, and mazelike grids, all painted red — decorate the slanting roof of the shelter. The concentric circle, as noted elsewhere (see site 8), represents the whirlwind and symbolizes the shaman's ability to concentrate power and/or to fly. The double diamond, configured as a short chain, represents the rattlesnake, one of the most potent of the supernatural spirits (see site 6). The meaning of the mazelike grids is unknown (see site 37), although they clearly represent entoptic patterns experienced during the initial stage of a shaman's trance. A black humanlike figure, presumably a spirit or shaman, is also present, as well as a set of small, finely drawn handprints and a series of fainter black pictographs. Finally, more petroglyphs are along the side of and behind the rock shelter.

Approximately in the middle of the shelter is a large rock with a number of cupules—small cuplike depressions made for ritual purposes on rock surfaces. These have been placed along a prominent corner of the rock as well as on its upper surface, creating a kind of serrated edge on the boulder. Although we lack ethnographic information about the making of cupules in southern California, it is likely that the grinding of these depressions was related to larger beliefs about the rock and the site itself; specifically, that the rock was the entrance to the supernatural world and grinding cupules into it allowed access to the supernatural power contained therein.

The presence of cupules at a rock art site is not unusual. There are many such examples in the Far West, although none are quite as dramatic as the one at Andreas Canyon. Our best hypothesis for the origin of the cupules is that they were made by nonshamans in what can best be termed "folk uses" of rock art sites, reflecting the

Photo 19: The Andreas Canyon rock shelter features pictographs on its ceiling, petroglyphs on the side, and cupules ground into rocks within the shelter.

sacred nature — and therefore supernatural power — of these places. Nonshamans sometimes visited sites when ill, to pray and make offerings in the hopes of recovery, which makes an important point: Although the motifs were considered sacred (see site 30), the spirituality of the site was not disembodied from its context. The site as a whole, and especially the rocks upon which the art was placed, were integral parts of the symbolism and sacredness of the rock art–producing tradition. And this, of course, emphasizes all the more that the sites should be treated carefully and with great respect.

▀▀▀▀ Visiting Site 14: The Indian Canyons are open every day of the week except during the summer, when they are open only on Fridays, Saturdays, and Sundays. There is an entrance fee to the Indian Canyons.

The Andreas Canyon pictograph site is in the Indian Canyons area of Palm Springs (see map 8). Head south on Palm Canyon Drive through town, continuing on South Palm Canyon Drive after it turns toward the east. You will reach the entrance to the Indian Canyons about three miles down South Palm Canyon Drive. After entering the reservation, continue south to the Andreas Canyon turn, which is marked with a sign. Head west to the Andreas Canyon parking lot. The rock art is located on a prominent ridge immediately adjacent the northeastern corner of the parking lot. A short dirt path leads up to the rock shelter. For information on visitation times, contact the Tribal Office of the Agua Caliente Band of Cahuilla Indians; see appendix for the phone number.

Map 8: Andreas Canyon.

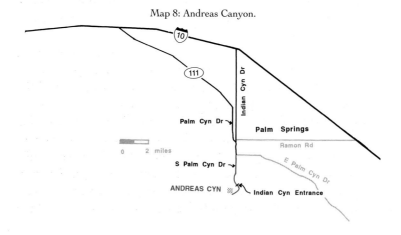

SITE 15
Fish Trap Petroglyphs
Coachella Valley, CA

Ancient Shores and Ancient Art

The Salton Sea is a familiar feature on the modern landscape of the Coachella Valley. Though perhaps considered a natural lake by many, the Army Corps of Engineers inadvertently created the "sea" in 1905. But that was not the first time the low-lying Coachella Valley filled with water; in the past 2,000 years it may have filled and drained three different times, with no assistance from the Corps, as the Colorado River changed courses—sometimes flowing into the valley and other times following its present course into the Gulf of California. Although 2,000 years sounds like a long time, from an archaeological perspective the episodes of filling and draining are relatively recent, and rather sudden, events.

Such dramatic fluctuations in the ancient body of water known as Lake Cahuilla obviously had important implications for prehistoric Native Americans. When full, the lake provided a bounty of foodstuffs, such as fish and water fowl, as well as other resources like reeds for basketry and shafts for arrows. As the level of the lake changed, turning lakefront property into desert fringe or vice versa, villages moved and the people had to adjust their hunting and gathering habits. The prehistoric picture that emerges is one in which the population experienced considerable flux as the natural environment changed.

Lasting side effects of Lake Cahuilla are deposits of travertine, a mineral precipitate that forms a relatively soft coating on flooded rocks. Travertine coats the boulders on the western slopes of the Coachella Valley, creating a kind of whitish band at the base of the mountains that marks the maximum height of the prehistoric lake. One area of travertine-coated boulders, near the present town of Valerie Jean, is a small petroglyph site with about a dozen motifs.

The rock art at this site consists of a large stick-figure human with splayed digits, both complex and simple entoptic patterns, and long sequences of short parallel lines or "tick marks" (see photo

Photo 20: At Fish Trap, petroglyphs are engraved on travertine-covered boulders. Note the line of vertical tick marks near the center of the photo; such sets of parallel lines were common entoptic patterns experienced during the first stage of a shaman's trance. A shaman might also use the marks to keep track of the number of spirit helpers he had received. The red paint is graffiti left behind by vandals. (Scale: most tick marks are about 6 inches long.)

20). Because this site lies *below* the ancient lake's most recent shoreline, which deposited travertine between about A.D. 900 and 1500, these petroglyphs must date to within the last 500 years and were apparently made by the Cahuilla.

As noted previously (see site 14), the Cahuilla tended to produce petroglyphs in the desert portions of their territory and pictographs in the wetter regions to the west, although some mixing occurred. Judging from the location and the nature of the motifs, the Fish Trap petroglyphs appear to represent vision quests rather than puberty initiations, and thus they portray the spirit world as seen by shamans.

The "tick marks" here are particularly interesting enigmas. They are common at many far-western North American sites, and many archaeologists have attempted to interpret them. We now have almost as many interpretations as we have tick marks, with suggestions ranging from menstrual cycles to hunting tallies to lunar/solar cycles. This, of course, illustrates the problem with inductive interpretations of rock art motifs: There are many potentially logical interpretations, but without guidelines for plausibility or additional evidence it is impossible to select one over the others.

Sets of parallel lines are one of the most common entoptic forms perceived during an altered state of consciousness. Given this fact, we should expect to see them at sites where shamanistic trances occurred. What they mean, however, is a different matter. Our only ethnographic account of tick marks comes from the Columbia Plateau rather than from California; it states that a shaman marked a rock with ticks every time his son, who was training to become a shaman, received another spirit helper. Whether the Fish Trap tick marks indicate counts of spirit helpers is, of course, entirely unknown.

A few hundred yards northwest of the petroglyphs, at about the same elevation along the base of the mountainside, is another Cahuilla archaeological site worth visiting. It consists of several stone fish traps made from horseshoe-shaped piles of cobbles and small boulders. These were used by the Cahuilla to capture fish in the tidal wash of the lake. Located near the edge of Lake Cahuilla's most recent shoreline, these particular fish traps are among the oldest preserved examples in the valley. As the lake level began to drop some 500 years ago, these fish traps were left high and dry and were replaced by others built at lower points to coincide with the lake level. More than a dozen different fish traps have been found and recorded by archaeologists at different elevations within the valley, charting the drying of the lake, which forced the Cahuilla to become true desert dwellers, at least until the Army Corps interceded in the early 1900s.

Visiting Site 15: The Fish Trap petroglyphs are located within a Riverside County park, although there are precisely zero signs, fences, buildings, park personnel, access roads, or other clues that such is the case. The site is about 10 miles south of Indio.

Take California Highway 86 through the Coachella Valley (see map 9). Turn west on 64th Street, at Valerie Jean, and follow it for

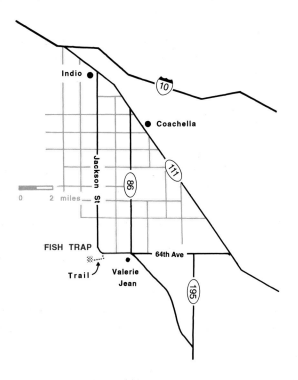

Map 9: Fish Trap.

about 2 miles to the L-shaped intersection of 64th and Jackson Streets. Park at this intersection. The petroglyphs are on the whitish, travertine-coated boulders on the hillside a few hundred yards west. A small trail runs along the south side and outside of the orchard between Jackson Street and the site, which will allow you to reach the site without trespassing in the orchard.

You will notice a lot of graffiti painted on the boulders; this, unfortunately, indicates that you are in the area of the petroglyphs. The fish traps lie a few hundred yards to the northwest, around a small point on the mountainside. Walking on the travertine-covered boulders destroys their fragile coating, so please tread very lightly and, as at all sites, do not climb or walk on the petroglyphs.

SITE 16
Morteros Village Pictographs
Anza-Borrego Desert State Park, CA

The Boys' Puberty Initiation

The Kumeyaay people, also known as the Tipai or Diegueño, left two small pictograph sites in the portion of their historical territory that now lies within Anza-Borrego Desert State Park. Kumeyaay territory stretched from the Colorado Desert to coastal San Diego County and extended down into northern Baja California, constituting a large portion of southwestern California. Influenced by their Yuman-speaking neighbors along the Colorado River, the Kumeyaay were somewhat transitional between the river groups and the core southwestern California cultures, such as the Gabrielino and Luiseño to their north.

Following southwestern Californian patterns, the Kumeyaay apparently produced pictographs in three contexts: First, as in all areas of the Far West, shamans made rock art following their vision quests to record the spirits and events they saw in the supernatural; second, young girls painted the spirit they acquired during the visions they received as part of their puberty initiations; and, finally, young boys also painted rock art to record the spirit helper they had obtained during their ordeals. Although we have no direct ethnographic evidence on this site, per se, it fits the general pattern described for southwestern Californian boys' initiation art.

One such characteristic of boys' sites is the use of black as opposed to red paint for the motifs. Black was the "male color" throughout much of the Far West, just as our culture associates blue with boys and pink with girls. For Native Americans, black represented east, the direction of dawn and its great supernatural power. The "female color" red, in contrast, was associated with west, the direction of the sunset and death. This does not mean, however, that only females could use red paint or that males had exclusive use of black; red was widely used by both sexes for face and body painting—but, in certain ritual contexts, it was the "female color."

101

One important ceremonial use of these gender colors involved the ritual body paint of the southwestern Californian shaman, known among some groups as the "Paha." During certain ceremonies, the Paha would paint himself, bilaterally, red and black, the two gender colors. Interestingly, the term *paha* also referred to the California racer, a snake that has both red and black color phases during its life span; southwestern California native groups incorrectly interpreted these phases as sex differences in the snakes. By painting himself with the two gender colors and by using the name of the snake that displayed dramatic visual sex differences, the southwestern California shaman mediated the opposition between the sexes; he represented a symbolic union that served to resolve the natural conflicts between males and females in society. In this same vein, rock art sites were referred to generically as "Paha's House," or the shaman's house.

The boys' puberty ceremony was an elaborate ritual intended to teach them the moral, religious, and philosophical precepts of their culture, to allow them to demonstrate their worthiness and inner strength by completing painful trials, and to give them an opportunity to acquire a spirit helper. It involved isolation in a ceremonial enclosure with one or more shamans and other adult males; instructions in dancing and singing; the creation of a sand painting, which was used to teach traditional cosmology to the boys; the administration of jimsonweed to induce the hallucinations from which spirit helpers could be obtained; and, sometimes, physical ordeals in the form of whippings with stinging nettles or rolling in a red harvester ant hill—both of which, not incidentally, can also induce hallucinations. The culminating event of the initiation, apparently, was a ritual race to a rock where the boys painted the spirit helper each had acquired. The boys who completed the initiation emerged as men.

The Morteros pictographs are on a large boulder within a Kumeyaay village that dates to the last 800 years. The pictographs are probably less than 800 years old, and perhaps only a few hundred years in age. One painted motif at the site—a set of pendant loops below a line (photo 21)—is very distinct, while other motifs on the panel are less distinct. Unfortunately, vandals armed with charcoal have added some recent graffiti to the panel.

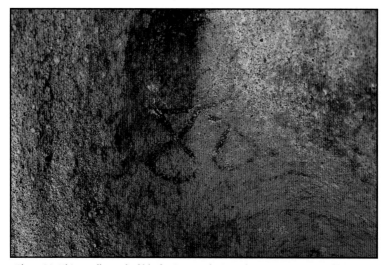

Photo 21: This small panel of black pictographs is at the Morteros site in the Anza-Borrego Desert. Although ethnographic evidence on the subject is limited, descriptions of ceremonies in southwestern California allow us to hypothesize that small panels of curvilinear motifs such as this may have been produced at the culmination of a boys' puberty ceremony, to depict the patterns of supernatural power the boys had seen during their visionary experiences. (Scale: the looplike motif is about 12 inches.)

The pictograph panel is located near the ground surface in a small declivity on the backside of a large boulder, relative to the trail through the site. A smaller boulder that sits adjacent to the rock art boulder has a series of cupules ground into it, further attesting to the ritual importance of this specific location within the village (see site 14).

▂▂▂ Visiting Site 16: Morteros Village may be visited in conjunction with the Little Blair Valley pictographs (site 17). A short hike is required. Both sites are in Anza-Borrego Desert State Park (see map 10).

Take road S-2 south toward Shelter Valley from California Highway 78. You will enter the park after traveling about 4.5 miles, approximately 0.3 miles south of the Stagecoach Trails RV Park and Store. A dirt road, marked with a sign to a historical marker, intersects S-2 from the southeast (left side of the road) about 1.6 miles farther south. Head southeast on this road for about 3.8 miles, following the signs to Morteros. (Note that all turns off the main track are marked.)

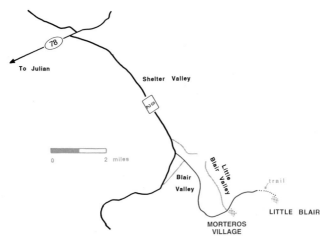

Map 10: Morteros and Little Blair Valley.

Southeast of the car park and up a short trail, Morteros Village is marked with a sign.

To find the pictographs, follow the designated trail roughly 300 yards to the large split boulder, behind which are a series of bedrock mortars, or "morteros." Walk back along the trail from the split boulder toward the car park for about 110 yards. You will see a large boulder roughly 10 feet high, about 15 yards from the trail on your left (northwest), with a smaller boulder about 4 feet high alongside, but not touching it, to the east. The smaller boulder has a series of cupules ground in it. Continue around to the backside of the large boulder. The pictographs are in a small panel near the ground. There is also a bedrock mortar on a horizontal rock nearby.

MAP 115
D-9

SITE 17
Little Blair Valley Pictographs
Anza-Borrego Desert State Park, CA

The Kumeyaay Girls' Ceremony

This rock art site in Anza-Borrego Desert State Park consists of two pictograph panels, together containing about two dozen motifs. Painted entirely in red, these motifs are predominately zigzags

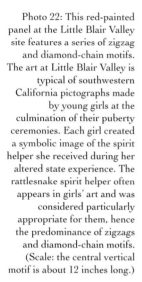

Photo 22: This red-painted panel at the Little Blair Valley site features a series of zigzag and diamond-chain motifs. The art at Little Blair Valley is typical of southwestern California pictographs made by young girls at the culmination of their puberty ceremonies. Each girl created a symbolic image of the spirit helper she received during her altered state experience. The rattlesnake spirit helper often appears in girls' art and was considered particularly appropriate for them, hence the predominance of zigzags and diamond-chain motifs. (Scale: the central vertical motif is about 12 inches long.)

and diamond chains (see photo 22), or some combination thereof; others include a sunlike motif, a dot pattern, and parallel lines, along with some very faded designs.

The red paintings at Little Blair Valley contrast nicely with the black pictographs at Morteros Village (site 16). Because of the gender distinction in the use of red and black paint, discussed previously, and the predominance of zigzag and diamond-chain motifs, we can infer that the Little Blair Valley pictographs resulted from Kumeyaay girls' puberty initiations. Most likely, the age of this site ranges from only a few hundred years to less than a thousand years old.

Analogous to the boys' ritual, the girls' ceremony marked their transition to womanhood, puberty, and eligibility for marriage; it was timed to correspond roughly with menarche, the first menstrual period. Unlike the boys' ceremony, however, the girls' initiation included a detailed enactment of the proper procedures and rites associated with giving birth, thus emphasizing the perceived primary role of Native American women in southwestern Californian society. Among these procedures, the girls had to lie in a shallow stone-warmed pit for three days, mimicking the ritual isolation and immobility they would practice following actual childbirth. To help them obtain a spirit helper, however, they ingested native tobacco prior to being incarcerated in the pit. As previously indicated — and contrary to our cultural perceptions of tobacco as a casual vice — Native Americans ate or smoked the ground leaves to induce an altered state of consciousness because the natural plant is in fact a potent hallucinogen.

The purpose of the girls' altered states was to experience a vision and thereby receive a spirit helper. Paramount among such helpers received by girls was the rattlesnake (see sites 10 and 21), the spirit seen in the supernatural world as an entoptic pattern: zigzag or diamond-shaped chain. Indeed, the importance of the rattlesnake as the most appropriate women's helper was so great that a successful initiation required the appearance of a supernatural snake.

The initiation culminated in the creation of pictographs. They ran a ritual race from their village to the designated rock art site, where elders (generally, a shaman) watched over the painting process. As noted earlier, the ostensible purpose of the race was to demonstrate health and vigor; the winner was thought most likely to live the longest among the group of initiates. As coincidence would have it, the winner of the last recorded girls' puberty ceremony in southwestern California, which occurred near Pauma in the 1890s, was the longest-lived of her female cohorts.

▰▰▰ **Visiting Site 17:** Little Blair Valley may be visited in conjunction with the Morteros Village pictographs (site 16). A short hike is required. Both sites are in Anza-Borrego Desert State Park (see map 10).

To reach the Little Blair Valley pictographs, follow the Little Blair Valley Road off S-2 to Morteros Village (see site 16 and map 10). From the Morteros Village car park, take the dirt road heading northeast, which is marked with a small "Pictographs" sign. You will encounter a car park, sign, and trailhead about 1.4 miles from Morteros Village.

The mile-long walk to the pictograph site takes you out of Little Blair Valley into Smuggler Cove, which is just over a low pass to the east. The trail begins by heading up a rocky wash and smoothes out to sand within a few hundred yards. The rock-lined trail dead-ends at a large isolated boulder on the right side of the path. Both of the painted panels are easily visible on this boulder, at about eye level.

SITE 18 MAP III

Corn Spring Petroglyphs F8
Desert Center, CA

Hallucinations on the Rock Face

The Corn Spring petroglyph site is in the Chuckwalla Mountains, in a transitional zone between the Desert Cahuilla and the Yumans of the Colorado River region. This important locale provided water along what was once a major east-west Native American trail, connecting Corn Spring with the Colorado River region to the east and the Coachella Valley to the west.

The Cahuilla were a large ethnolinguistic group whose territory ranged from the desert floor north of the Salton Sea, through the mountains above Palm Springs, to the climatically more moderate area around modern Riverside. Those who resided in the desert region necessarily followed a lifeway similar to their desert-dwelling Numic neighbors to the north. Unlike the Numic, however, at least some of the Cahuilla bands practiced girls' puberty ceremonies, during which pictographs were painted. Shamans also created rock art to portray the spirits of their visionary images. Although we have no direct ethnographic evidence about the Corn Spring site, it most likely served as a vision quest site for shamans. Following a common pattern for desert rock art sites, it is associated with a permanent spring as well as the remnants of a prehistoric village.

Petroglyphs at this site are along both sides of a wide desert wash; some on the north, immediately adjacent to the road, are marked by signs. As is typical of most sites, almost all of the petroglyphs here are entoptic forms from the initial stages of a shaman's altered state. But there are also some stick-figure humans portrayed, including some with crownlike headdresses. The petroglyphs at this site appear to date primarily from the last few thousand years, with many probably made in the last 1,000 years.

Some of the most elaborate petroglyphs are on the south side of the wash. One of these, near the western end of the site, displays a good example of how a rock's natural features are sometimes incorporated into a rock art image (photo 23). In this case, a linear quartz vein anchors a series of half-circles or inverted Vs pecked along one side, creating a kind of fringed line that is itself probably a complex entoptic form.

Photo 23: A long, vertical quartz vein at the Corn Spring site has a series of loops engraved along its site. This motif exemplifies the common practice of incorporating natural features of the rock face into the art. While in his trance, the shaman stared at the rock face and construed the natural features of the rock face as parts of his visionary experience. Upon emerging from his altered state, he pecked his design of power on the rock panel. (Scale: quartz vein is about 5 feet long.)

The incorporation of a rock's natural features in the art is relatively common and appears to reflect two factors: First, ethnographic evidence suggests that, at certain sites, shaman initiates conducted their vision quests in part by staring at the rock face. Their visionary hallucinations sometimes then involved construing the natural features of the rock face into images, which they subsequently "traced" onto the panel. Second, the rocks were believed numinous, or inhabited by spirits. The spirits seen in the shaman's trance, including those that appeared in geometric or entoptic form, and were then painted or pecked on the rock face, were believed to reside inside the rock. The rock art thus portrayed what was thought to exist immediately on the other side of the rock surface, in the sacred realm.

Visiting Site 18: Corn Spring is adjacent to a developed campground that is easily accessible by a good dirt road from Interstate 10. Take the Corn Spring exit about 9 miles east of Desert Center and proceed to the south side of the freeway (see map 11). Follow the frontage road heading east for 0.6 miles until you reach a dirt road marked with a sign to Corn Spring. Turn right (south) down this road. The site is located about 7 miles from the frontage road, immediately before the campground, and is marked by signs. Contact the BLM Palm Springs Resource Area Office (see appendix) for information on the road or campground.

Map 11: Corn Spring.

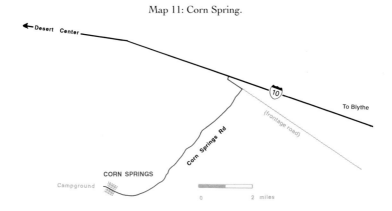

MAP 112
D-2

Visiting the Sites

SITE 19
McCoy Spring Petroglyphs
Blythe, CA

Ancient Art, Modern Graffiti

McCoy Spring is the largest rock art site in southeastern California. It requires a difficult hike—more than ten miles round-trip—but the effort is worthwhile for those who are capable and well prepared. Although the exact boundaries between different Native American groups are ambiguous, McCoy Spring most likely fell within Yuman (Mojave and Quechan) territory, and thus follows the general patterns found along the Colorado River region proper, to the east.

McCoy Spring has suffered considerably from vandalism. The site was used as a stagecoach watering stop after the 1880s, and travelers pecked names, initials, dates, and figures into an unusually large number of boulders at the site. Although this graffiti has certainly degraded the site, we can use the dated engravings to help calibrate the time required for the onset of revarnishing, thus allowing for a better understanding of the processes that allow us to date petroglyphs chronometrically. The historical graffiti, while we decry its presence, has become a kind of archaeological record in its own right that helps us better understand the Native American art that surrounds it.

The graffiti notwithstanding, McCoy Spring is a large, spectacular, and interesting site. Riverside archaeologist Dan McCarthy has mapped over two thousand petroglyph boulders here, many of which contain multiple engravings. There are also rock rings scattered in and among the rock art panels that served as foundations for brush huts, indicating that people lived here for at least part of the year. This is not surprising, since McCoy Spring is the only water source in the region. McCarthy has also identified a series of ancient trails radiating from the site area, which apparently served as a kind of hub for hunting and gathering parties.

Many of the motifs at the site are heavily revarnished, indicating that some of the art may be 5,000 years or more in age, although more

recent Native American petroglyphs are also present. This conti-
nuity in ritual activity at a single locale on the landscape suggests
that Native American religious traditions were, at least at some
level, very long-lived and conservative in nature. For example, the
association between rock art and springs clearly goes back many
centuries, judging from the evidence at this site. The belief that
springs served as entrances to the supernatural may be very ancient. But all activities in the desert were constrained by access to
water—even shamans had to drink—so we cannot be certain
whether the continuous use of this site is tied entirely to symbolism
or whether it is due, at least partly, to the physiological need for
water.

You will find petroglyphs on both sides of a deep, dry wash at
this site. Some are on small boulders near ground level on the flat
terraces above this major arroyo, as well as in the smaller tributary
washes to the sides. A large broken boulder in the middle of the
wash on the east side of the site, which truncates the old road,
contains a couple of interesting petroglyph panels (photo 24). It
includes a conflation of a human and a rattlesnake—that is, a human supernaturally transforming into his spirit helper, the snake.
In this case, the sinuous snake emerges from a crack in the rock
and transmogrifies into a stick-figure human form. The supernatural world was believed to lie on the far side of rocks—that is, within
the earth—throughout far-western North America. It was further
believed that individuals entered the sacred realm when cracks
opened in rocks, revealing tunnels that led to the spirit world. Animal species that moved in and out of cracks in rocks, such as lizards and snakes in particular, were therefore believed to be strongly
associated with the supernatural. Lizards, for example, were
thought to be messengers between the natural and supernatural
realms. The petroglyph on this large boulder, however, portrays a
snake-human emerging from the sacred world.

Other notable petroglyphs on this same broken boulder include
two stick-figure humans engraved with detailed fingers and toes.
Referred to as "digitate anthropomorphs," they are characteristic
of Yuman rock art. Unfortunately, modern vandals have also left
their mark, such as the very well-drawn bird motif on the same
panel and a finely chiseled bighorn sheep elsewhere on the site.

Photo 24: A petroglyph panel at McCoy Spring. The "bird" at the upper right is an example of the historical graffiti common at this site, which once served as a stagecoach stop. More important is the human-snake conflation in the image on the left, portraying a Rattlesnake Shaman transforming back into a person from his supernatural alter ego, the snake, as he emerges from a crack—a portal to the sacred realm—in the rock face. (Scale: human-snake figure is about 18 inches tall.)

I have implied above that the petroglyphs at McCoy Spring, as with those at other sites in this guide, represent visions of the supernatural world. While this is correct, if this site was made by Yuman-speaking groups, as we assume, the art nonetheless did not have meanings that are fully equivalent to the other California sites discussed so far. This is because the Yuman speakers along the Colorado River maintained different beliefs about the supernatural and had slightly different rock art–producing traditions that involved two different ritual activities.

First, as in all other parts of southern California and Nevada, shamans made rock art to portray their visionary experiences. The Colorado River Yuman-speaking peoples, however, believed that when a shaman entered the supernatural he reexperienced the mythic creation of the world, visiting the creator deity, Mastamho. Thus, the Yuman shaman entered a kind of supernatural world best termed "mythic time-space." This was different from other Californian and Great Basin groups. For them, mythology was very distinct from the sacred realm, which paralleled the mundane temporally and chronologically.

The primary, but not sole, site of Yuman shamans' supernatural-mythic experiences was Spirit Mountain, or Avikwa'ame, north of Needles (see site 23). But they also apparently entered the sacred realm at other places where they undertook the same or similar kinds of supernatural activities characteristic of shamans in general. For example, Yuman speakers had Rattlesnake Shamans who could handle snakes, cure bites, transform into snake spirits, and so on, like many other Native American groups. They also had sorcerers, or evil shamans, who could inflict "ghost sickness" on people. One account relates that a Mojave sorcerer made petroglyphs of Euro-Americans in an effort to bewitch them and, presumably, cause them to leave the Colorado River region.

Since the Yuman shamans generally received their powers from Mastamho, their art does not primarily depict spirit helpers; instead, it portrays spirits of the mythic world, the shaman in that realm, and, especially, highly condensed and very potent images of mythic events. In a real sense, then, the rock art at McCoy Spring likely portrays Yuman mythology, but not in any narrative form. Since the perception of these myths occurred during an altered state, the entoptic motifs common at the site served as an abstract sign of individual visions of the mythic past. A parallel to this kind of graphic imagery occurs in our own culture, where many Christians display a highly stylized outline of a fish to proclaim their faith. For them, the fish symbolizes a certain sacred "myth"—Christ multiplying the loaves and the fishes—but not in any narrative fashion; rather, it signifies a larger concept of Christ as the fisher of human souls.

In addition to shamanic rock art, Yuman-speaking groups also created ritual rock art during boys' puberty ceremonies, at which time their nasal septums were pierced for nose ornaments. The boys were shown rock art panels prior to the ceremony and might have repecked selected images they thought would be important to them. Although we only have two ethnographic descriptions of locations where the nasal-piercing ceremony occurred and where rock art was made, neither of which is McCoy Spring, it is also possible that some of the art at this site may have been made by male puberty initiates. As with the shamans, however, the boys too "dreamed" of the mythic past and portrayed it in their art, but, again, in a condensed, nonnarrative form.

Probably the largest concentration of petroglyphs at McCoy Spring is around the spring itself, across the arroyo and north of the fence at the site area. The majority of the engravings are entoptic patterns (photo 25), as is the case with all sites, although some human figures are also present. Although Dan McCarthy has intensively recorded this site, he has found only two bighorn sheep petroglyphs. Additional petroglyphs are down the canyon, to the west of the fence.

ΛΛΛ Visiting Site 19: McCoy Spring lies within a designated wilderness area. You must hike the final 5.1 miles to McCoy Spring, and you will need a high-clearance, four-wheel-drive vehicle to reach the trail that leads into the site.

Photo 25: A relatively heavily revarnished petroglyph from McCoy Spring, the largest site in the Colorado Desert region. Like most petroglyphs at this site, this is an entoptic pattern seen by a shaman during a trance. (Scale: motif is about 18 inches long.)

Caution: Because of harsh desert conditions, only experienced, physically fit, properly equipped hikers should visit McCoy Spring, and then only during relatively cool weather. Temperatures in the Colorado Desert can be treacherous from late spring through early fall. There is no water at the spring, so be sure to carry enough for the entire trip in addition to the length of your stay. Although it is possible to hike in and out in one day, an overnight visit will allow more time to see the art and enjoy the experience.

To reach McCoy Spring, follow I-10 to the Wiley's Well exit, about 14 miles west of Blythe (see map 12). The road to the wilderness area boundary is a hard, slow drive. While it does not require any dangerous four-wheel-drive maneuvers, the road crosses numerous small washes, requiring stop-and-go driving for most of the distance.

Map 12: McCoy Spring.

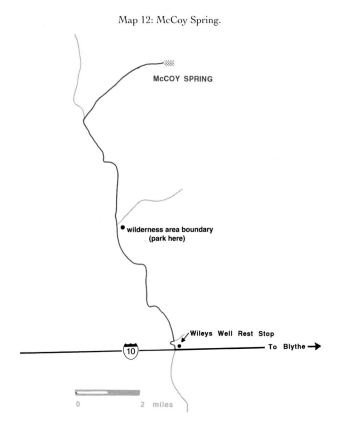

McCOY SPRING

• wilderness area boundary
(park here)

Wileys Well Rest Stop

10   To Blythe →

0   2 miles

Head north from the freeway toward the Wiley's Well rest stop. Instead of entering the rest stop, which will be on your right, proceed straight ahead from the freeway exit onto a dirt road immediately west of the stop. Within a few yards you will encounter a high dirt berm. Drive west (left) to the end of this berm and then double back toward the east-northeast to a dirt road following an east-west–running power line. You will encounter a dirt road heading north from the power-line road, approximately due north of the rest stop. Take this north-trending dirt road, which will lead you past an abandoned homestead and will begin to head northwest.

Follow this road north-northwest. About 5.2 miles north of the power line you will cross a dirt road heading off to the right and a wilderness area boundary sign. Park here and continue on foot straight up the road. About 2.7 miles from your car, a barely visible two-track road will cross your path. Do not turn; continue following the road you came in on about 0.3 miles farther, at which point you should see a road on your right, heading to the east-northeast. Follow the main track of this new road approximately 2.1 miles as it leads toward the McCoy Mountains. The road will dead-end at a large brown aluminum fence. The main portion of the McCoy Spring site starts at this fence, and heads east on both sides of the arroyo to your left.

All visitors to McCoy Spring should consult and carry two USGS maps—the Hopkins Well and the McCoy Spring 7.5-minute quadrangles—in addition to this book. For more information, contact the BLM Palm Springs Resource Area Office (see appendix).

MAP 112
F2

SITE 20
Mule Canyon Intaglios
Blythe, CA

Dance Circles and Tank Tracks

The Colorado River Valley and its immediate environs contain an important series of intaglios, or earth figures. Native Americans created them by clearing away cobbles, rocks, and pebbles from the desert pavement, leaving a recessed area of light-colored earth surrounded by darker ground. The Mule Canyon site, southwest of Blythe, contains one possible example of these earth figures. Although the site is worth seeing in its own right, it is directly on the way to the Mule Tank petroglyphs (site 21) and can easily be visited en route.

The intaglios are believed to have been made by Yuman-speaking groups, such as the Mojave and Quechan, who resided along the Colorado River. Unlike petroglyphs and pictographs, which reflected shamans' altered states of consciousness, the intaglios were apparently used in group ceremonies, including pilgrimages to the sacred mountain, Avikwa'ame. The ceremonies involved long recitations of song cycles and dancing, sometimes continuing for an entire night. The songs pertained to different myths and mythic events, pointing to the close relationship between mythology and ritual among these Yuman groups, which did not hold for many other Native American groups in California and the Great Basin.

I have emphasized above that the Mule Canyon site may or may not be Native American in origin. One alternative interpretation is

Photo 26: One of the intaglios, or earth figures, at Mule Canyon. Although Native American groups along the Colorado River made intaglios, archaeologists are not certain whether they made these horseshoe-shaped patterns of concentric circles scraped into the desert pavement. (Scale: each circle is about 36 inches in diameter.)

that the cleared patterns at this locale, which consist of a series of horseshoe-shaped clusters of about ten circles apiece (photo 26), may have resulted from ground surface disturbance during training exercises by General George Patton's troops prior to World War II. Four circumstances support this possibility: First, it is known that Patton conducted war exercises in this area, and foxholes and shell casings have been found in the immediate vicinity of the site; second, there are no known Native American geoglyphs that are similar in form to these; third, the cleared circles lack the small berms of rocks and pebbles around their margins that tend to characterize all other intaglios; and, fourth, an early archaeologist, Malcolm Rogers, recorded a trail and dance circle at this site before World War II, but made no note of these horseshoe-shaped circle clusters, suggesting that they were not present before the war.

Although we may never known for certain whether or not these patterns were created for religious or military purposes, there is also evidence supporting a Native American origin for these ground features. First, both Russ Kaldenberg and Boma Johnson, archaeologists with extensive experience in this region, have interviewed former members of Patton's troops. None of their informants recall having been at this specific locale, and none can cite any kind of military equipment or materiel that could have been responsible for these concentrations of circles. Kaldenberg has also measured the circles and compared their diameters to that of a 55-gallon drum, on the possibility that the ground circles may represent clusters of drums, but in all cases the circles in the pavement are about twice the circumference of standard 55-gallon drums. Second, there are no other examples of such concentrations of circles at other known war-games sites in the region, including ones that have disturbed desert pavements, further discounting the likelihood that they are military in origin. Third, the site also contains two features that are widely acknowledged as Native American in origin, as implied above: an aboriginal trail, which cuts through the site, and a dance circle, which is partly bisected by the trail. The presence of the dance circle at this site, in particular, indicates that the locale was used for ceremonial purposes, thereby increasing the likelihood that the clusters of circles are aboriginal as well.

Aboriginal trails and dance circles were created in a manner similar to the intaglios, though perhaps inadvertently: The pounding of feet pressed the rocks of the desert pavement into the soil and/or resulted in the rocks and pebbles being pushed aside, again creating a negative pattern in the dark desert pavement. The trail at the site is clearly visible, cutting through the long axis of the fenced-off site area. The dance circle is relatively hard to see, requiring good lighting to make out. It is partly bisected by the trail and the fence around the site.

The concentrations of circles, in contrast, are clearly visible. At least one of these has within it a series of small pebbles beginning to emerge from the soil, presumably due to standard soil processes which, over time, cause rocks to be upthrust slightly. If this interpretation is correct, it suggests that one or more of the concentrations of circles may be older than the others, further supporting a Native American origin for these intaglios.

Since the origin of these ground circles is controversial, we obviously do not known their age. Chronometric varnish dating of intaglios elsewhere along the Colorado River region suggests that some of them may be as much as 2,900 years old. Perhaps more importantly, it appears that some of them are still being used by Native Americans for ceremonies today.

Visiting Site 20: Four-wheel drive is recommended. You may visit Mule Canyon intaglios in conjunction with Mule Tank petroglyphs (site 21).

The Mule Canyon earth figures are reached by taking the Mesa Verde off-ramp from Interstate 10, on the west side of Blythe (map 13). Head south from the freeway for a short distance to the termination of the pavement at a T-shaped intersection with Blythe Way. Turn right (west) down Blythe Way (dirt) and follow it for about 1.1 miles. At that distance you will encounter a dirt road heading south on your left, paralleling a small power line and a jojoba bean field. Head south down this road for approximately 2.0 miles. Turn right at that point, heading west for 0.5 miles. Here you will encounter another north-south road, paralleling the first you took off Blythe Way. Turn left on this second road, continuing south down it for only 0.3 miles, at which point you will encounter a dirt road angling in from the southwest. Head southwest down it. You will cross a major

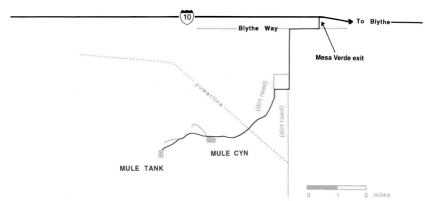

Map 13: Mule Canyon and Mule Tank.

power line and dirt road at about 1.5 miles from the last turn. The geoglyphs, which are in a signed, fenced enclosure, are about 1.0 miles beyond the power line.

The desert is crisscrossed by dirt roads in this area, and there are a number of alternative routes to get to the geoglyph site. A number of the dirt roads, including the power-line road one mile before the site, run through deep sand, and are unpassable except with four-wheel drive, so be careful about following alternative routes.

For information on road conditions, contact the BLM Palm Springs Resource Area Office (see appendix for details).

MAP 112
F2

SITE 21
Mule Tank Petroglyphs
Blythe, CA

The Female Shaman

Located about two miles beyond the Mule Canyon earth figures, the Mule Tank petroglyphs consist of a concentration of engravings found at a deep tank—in this case, a very low spot in a seasonal arroyo that retains water in a pool for a portion of the year. The Native American group that controlled Mule Tank historically as well as prehistorically is unclear. As noted in reference to McCoy Spring (see site 19), the Colorado River Valley and its immediate periphery were generally occupied by Yuman speakers, such as the Mojave. During the historic period, however, the Chemehuevi, a Southern Paiute band, also moved onto the river, in the vicinity of Blythe. It is possible that Mule Tank was, at one point, part of a Yuman group's territory and that, at another time, it fell within the Southern Paiute domain.

This uncertainty notwithstanding, the Mule Tank petroglyphs tend to fit, like those of McCoy Spring, the style of Colorado River region rock art: deep and widely engraved geometric petroglyphs showing evidence of repecking. As with most sites, entoptic forms dominate, thus representing the initial stages of a shaman's trance. The few representational forms present tend to be human figures, often stick figure and "digitate" or headless in form, as is common in this region.

The Mule Tank petroglyphs are found on both sides of a sinuous wash, at the top of which is the tank itself. Like McCoy Spring (site 19), and though not a spring per se, Mule Tank is the only source of water in the immediate region. Archaeologist Dan McCarthy has recorded an aboriginal trail that ran from it to Corn Spring (site 18), the next water source in the east-west trek across this section of southeastern California.

To the north of the tank, and about 20 feet above it, there is an unusual panel displaying a small human figure and a diamond-chain pattern (photo 27). This panel is notable for two reasons: It contains one of the relatively rare examples of representational images at the site (the human figure), and this human figure appears to be female. Based on general conventions in rock art representation across the West, female figures are typically represented by the depiction of pendant labia and/or a characteristic bottle-shaped body, as opposed to straight up and down. In this case, pronounced pendant labia are evident between the legs.

In California and the Great Basin, female rock art figures are rare. In the Coso Range (see site 1), for example, which has the largest concentration of "patterned-body anthropomorphs" in the Far West, I have recorded 400 of these human figures, and only about a half-dozen are female—less than 2 percent of the total. Female human figures are rare in the art because females were

Photo 27: A stick-figure human petroglyph. Note the depiction of what appear to be pendant labia on this figure, suggesting that it is a portrayal of a female. Clear depictions of females in the rock art of far-western North America are rare, reflecting the fact that women rarely became shamans because of the belief that menstrual blood was hostile to supernatural power. (Scale: human figure is about 12 inches.)

rarely shamans, as I have noted earlier. That is, while women could be shamans, in practice they seldom were. In a precisely similar vein, there is nothing in our Constitution or laws preventing a woman from becoming president of the United States yet, as we all know, in over 200 years such has not occurred because of cultural biases and other factors.

Among southern Californian and Great Basin Native American cases, the ostensible primary impediment to female shamanism was menstruation: Menstrual blood was inimical to supernatural power. During their periods women were, accordingly, barred from attending, let alone participating in, any kind of rituals; and should a drop of menstrual blood fall on a male hunter or his bow and arrows, it would ruin his luck with game (see site 10). The result was that women were more or less effectively barred from becoming shamans until they had reached menopause. Even then, a female shaman was an oddity. Among many groups, they were necessarily believed to be sorceresses, or evil shamans.

It is important to note, as an aside, that shamans were primarily women in some far-western North American groups, particularly in northwestern California, so this gender exclusion was not universal. Furthermore, on modern reservations there now seems to be a greater tendency for women to become shamans, suggesting that the traditional gender roles have changed.

The figure at Mule Tank, then, most likely represents a rare example of a female shaman. Since supernatural power was thought to independently "select" who would receive it, even though it usually ran in family lines, and since it manifested itself in dreams and visions, an occasional woman might have spontaneous visions and thus be earmarked as a shaman, going against the "normal" cultural pattern. Appropriately enough, the geometric motif alongside this human figure is a weblike pattern of connected diamond chains. As emphasized many times above, diamond chains were stylized rattlesnake motifs, modeled after the scale pattern on the western diamondback. Moreover, although there certainly were plenty of male Snake Shamans throughout the Far West, rattlesnake motifs were nonetheless "female" gender designs (see site 10). For example, boys' and girls' cradleboard hoods were decorated with a woven or embroidered pattern which, on one level at least, served to distinguish male from female babies, just as we use blue or pink clothing and accessories to differentiate our babies. The zigzag and the diamond chain were precisely those patterns used on the cradleboards of baby girls (see sites 17, 34, 35, and 36). In this case the gender of this human figure, then, is further emphasized by the motif associated with female gender, placed alongside it.

▰▰▰ **Visiting Site 21:** Four-wheel drive is required. You may visit Mule Tank petroglyphs in conjunction with Mule Canyon intaglios (site 20).

Mule Tank is reached by traveling another 1.8 miles west-southwest beyond the Mule Canyon intaglio site (see site 20 and map 13). From the western side of the fence at the Mule Canyon site, take the dirt road heading west. At about 1.0 miles from the west side of the Mule Canyon fence you will drop into a small arroyo. The road branches here, with one fork heading northwest to cross a large wash on your north (right), but continue straight ahead at this fork, staying on the south side of the large wash and winding along the foot of Mule Mountain. Continue beyond this first fork for approximately 0.8 miles, which will bring you around to the northwest side of Mule Mountain. The site extends from a fence on your left up the wash to the tank.

The BLM Palm Springs Resource Area Office may be checked for information on the road (see appendix for details).

MAP 112
D-3

SITE 22
Blythe Intaglios
Blythe, CA

Creation Ceremonies and Ritual Pilgrimages

The Colorado River intaglios or earth figures were made by Yuman-speaking peoples. As noted previously, they were created by scraping a design into the desert pavement of cobbles, rocks, and pebbles that covers the terraces above the rivers, exposing the lighter soils underneath. The giant earth figures north of Blythe, particularly good examples of Colorado River region intaglios, include three large humans, a feline, and a concentric circle and a spiral (photo 28). Elsewhere along the river terraces there are additional human and animal designs, including rattlesnakes, mountain lions, a lizard, and even a horse, and stars and other geometric patterns.

A number of the human-figure intaglios along the river are said to depict mythic actors and events, particularly the creator-deity Mastamho and his evil brother Kaatar, with the mountain lion representing Mastamho's creation helper. Research by Boma Johnson, an archaeologist living in Yuma, Arizona, indicates that a number of the intaglio sites were ritual stops along a Yuman pilgrimage route. The route ran from Pilot Knob, or Avikwal, the spirit house

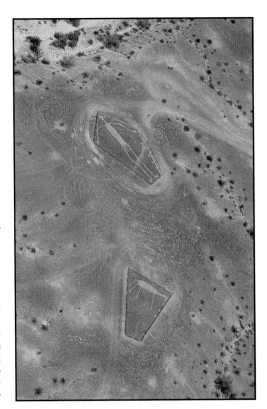

Photo 28: Aerial view of humanlike and zoomorphic (possibly a mountain lion) earth figures. These and similar intaglios were scraped into the desert pavement and used by native peoples along the Colorado River on ritual pilgrimages and to call the Creator. Typically, the images depict mythic actors and events. (Scale: human figure is about 60 feet long.)
—Courtesy of Harry Casey

where the dead dwell at the southern end of the river, to Avikwa'ame, or Spirit Mountain, where the Earth was created, in the north (see site 23). This pilgrimage was intended to honor the creation, and ritually retraced the path of Mastamho in his mythic adventures. At the various ceremonial stops along this route, an officiating shaman would instruct the participants in mythic history and ritually reenact mythic events, thereby preparing the participants spiritually for their experience at Spirit Mountain.

Although the Blythe giant figures are located alongside the ceremonial route, they have not been directly associated with the pilgrimage, per se. Boma Johnson, nonetheless, has recorded two slightly different accounts of the creation of these particular intaglios from the Mojave and Quechan that illuminate the specific ritual use to which they were put, and also indicate that they were important to more than one cultural group. According to the Mojave, at least one of the human figures represents the creator-deity Mastamho and was created by them, with the god's help, to seek his aid. They did this by dancing around the figure for three days and nights, calling on their creator. Notably, a dance circle is associated with one of these large human figures. The Quechan version, similarly, indicates that giant earth figures were created to celebrate the aid Mastamho rendered to humankind at the time of the creation.

Judging from these accounts, it is most likely that the Blythe earth figures were created in ceremonies intended to call on Mastamho, the mythic creator-deity, to solicit his assistance. This

Map 14: Blythe intaglios.

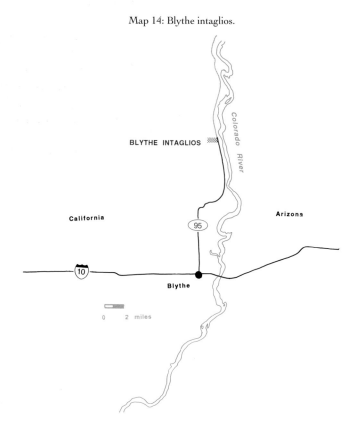

practice appears to be continuing to this day, in that the precise outlines of these figures seem to change from year to year. This emphasizes, all the more, the need to treat these figures with great care and respect. Unfortunately, prior to the fencing of the figures, circular motorcycle tracks cut into them, diminishing their integrity but, hopefully, not their spirituality.

Based on the existence of a horse intaglio among the corpus of Colorado River earth figures, we can infer that at least some of these intaglios were made in the historical period. Chronometric varnish dating also indicates that at least some of them are almost 3,000 years old. But more to the point, varnish dates have been obtained specifically on three of the Blythe intaglios: the two human figures and the mountain lion. Each of these was found to date to approximately 1,100 years ago, suggesting that the first ritual activity at this site occurred over a millennium ago.

Visiting Site 22: You can reach the Blythe intaglios by driving about fifteen miles north on US Highway 95 from Interstate 10 at Blythe (see map 14). The earth figures are located up a short dirt road on the western side of the highway, and a BLM sign on the highway will direct you to them. They are currently fenced off to prevent people from driving and walking on them. Though this is a necessary precaution, it makes them difficult to photograph.

Check with the BLM Palm Springs Resource Area Office (see appendix) for information about this site.

MAP 97
G-7

Southern Nevada Region

SITE 23
Grapevine Canyon Petroglyphs
Laughlin, NV

Spirit Mountain and the Patterns of Creation

For the Yuman-speaking inhabitants of the Colorado River region, no location was more sacred than Avikwa'ame, or Spirit Mountain, which we call Newberry Peak, in southernmost Nevada. According to the Mojave creation myth, the oldest spirit was Matavilya, made from the mating of Earth and Sky. Matavilya had two sons, Mastamho and Kaatar, and a daughter, Frog. Matavilya committed an unwitting indecency that offended his daughter, who then killed him. Mastamho directed the cremation and mourning ceremony for his father and, when completed, strode up the Colorado River Valley. When he got to the top Mastamho created the river by plunging a cane of breath and spittle into the earth, allowing the river to pour forth. Riding a canoe down the waters to the ocean, he created the wide river bottom by twisting and turning the boat. He returned from the ocean with his people, the Mojave, taking them in his arms to the northern end of Mojave country. There he piled up earth, creating the mountain Avikwa'ame, and built himself a house on it. There too Mastamho plotted the death of Sky-Rattlesnake, an evil spirit and the source of dark powers. Mastamho killed Sky-Rattlesnake by cutting off his head, with his spilt blood becoming noxious insects. Mastamho then gave land to the different tribes and taught them to farm. Finally, Mastamho turned himself into a fish-eagle and flew off into oblivion.

The importance of this cosmogenic myth to rock art is twofold. Known as the "Shaman's Tale," it was precisely this myth that the shaman "dreamed" to obtain his supernatural powers: In Yuman fashion, the shaman was believed to reexperience and witness these mythic events of creation in the supernatural world and, from them,

128

Photo 29: A complex panel of petroglyphs located at the foot of *Avikwa'ame*, or Spirit Mountain — the Yuman place of creation. Shamans went to Spirit Mountain during their trances to reexperience the mythic creation of the world. (Scale: panel is about 4 feet high.)

obtain his power. As the anthropologist Alfred Kroeber wrote in 1925, "It is of [Mastamho's] house that shamans dream, for here their shadows were as little boys in the face of Mastamho, and received from him their ordained powers, confirmed by tests on the spot." It is at the foot of Spirit Mountain that the important Grapevine Canyon petroglyph site is located (photo 29); that is, Grapevine Canyon is Mastamho's House, where the Mojave shaman went to witness, in his dreams, the creation of the world.

Grapevine Canyon is the biggest petroglyph site in southern Nevada. As at other Colorado River locales, representational or figurative motifs are rare at the site; they are heavily outnumbered by entoptic designs, many of which are deeply and widely engraved (photo 30). The depth and width of these engravings reflect the fact that Yuman petroglyphs were studied and examined by initiates, and sometimes repecked by them. Still, considerable variation in the degree of revarnishing is evident at the site, indicating that this locale was used by Yuman shamans probably for many thousands of years. Recent accounts, on the other end of the time scale, tell us that it continued to be used into this century.

Many of the entoptic engravings at the site are complex combinations of geometric forms; enclosed grid and netlike patterns and shieldlike motifs are particularly common. On some panels the dense overlay of engravings itself creates an almost uninterpretable, complex geometric motif; individual engravings can only be identified through careful examination of the dense palimpsest of petroglyphs. There are some familiar motif types present at the site, however, such as occasional human figures (including a few "patterned-body anthropomorphs"), and even a few panels of bighorn sheep. Whether these represent Yuman petroglyphs or instead were created by non-Yuman shamans is unknown. Avikwa'ame was widely renowned as a place of great supernatural power, so it is entirely possible that shamans from different cultures came here for their own types of vision quests.

As implied above, the petroglyphs at Grapevine Canyon apparently represent shamans' visionary images of Mastamho and the creation of the world. Clearly, though, these images do not comprise narrative depictions of the different events in the creation myth, as I have noted in a similar vein concerning the petroglyphs at McCoy Spring (site 19). Instead, they are signs of the mythic event and, in this fashion, perfectly parallel the verbal descriptions of the creation that shamans included in their mythic song cycles. As anthropologist Alfred Kroeber noted in 1957, although the Mojave may dream a mythic event or series of animal spirits pertaining to a song cycle, they do not say that they dream the events or spirits, per se, but instead claim that their vision is of the pattern of the myth or spirit. Just as their graphic depictions of the creation of the world were expressed simply as entoptic designs, so too the verbal accounts in their song cycles were often simply a combination of a few words or lines that lacked any obvious narrative information. In both cases, this resulted because the actors and events in the myth were known by everyone, and the songs and petroglyph motifs were just the idiosyncratic signs of the mythic events and required no detailed illustration or recitation.

ΛΛΛ Visiting Site 23: Grapevine Canyon is easily located just outside of Laughlin, Nevada. To get to it, head west out of Laughlin on Nevada Highway 163 (see map 15). About 4.6 miles west of the

Colorado River you will encounter Christmas Tree Pass Road on your right (north) side, which is well marked with a sign. Turn north on this dirt road and proceed for about 1.9 miles, to the first turnoff on your left (west). This will lead you into a linear parking area about 200 yards off the road. Park here and walk along a large wash to the site, which is located about 300 yards west of the parking area on the rocks on either side of the wash, at a spot where it narrows into a rocky canyon. The site is easy to see because it has, perhaps unfortunately, become a popular local picnic area and there are usually a half-dozen or more cars in the parking area. The crowds notwithstanding, visiting this site is well worth the effort.

Grapevine Canyon falls within the Lake Mead National Recreation Area. For information on the roads and the site, contact the Alan Bible Visitor Center at the junction of US 93 and Lakeshore Road (see appendix for phone number).

Map 15: Grapevine Canyon.

Photo 30: Yuman shamans made petroglyphs to portray their reexperiencing of the mythic creation of the world. Instead of depicting a narrative illustration of the event, however, they show the essence or pattern of the creation as revealed by the entoptic patterns they saw during their trances.

SITE 24
Atlatl Rock Petroglyphs
Valley of Fire State Park, NV

The Power of Weapons and the Power of Place

Two very impressive rock art sites are located within Valley of Fire State Park, northeast of Las Vegas. Both provide an interesting contrast with the other sites discussed in this guide because the prehistory of the Valley of Fire region differed in some significant ways from that of the rest of the Far West. During the period from approximately A.D. 1 to about 1200, the Valley of Fire region was occupied by pithouse dwellers and eventually by farmers living in small permanent villages. They were similar in many ways to the prehistoric Anasazi or Pueblo peoples of Arizona and New Mexico, rather than to the archaic hunter-gatherer cultures typical of the larger Great Basin. This period corresponds to a more favorable, wetter climatic regime that essentially allowed rainfall agriculture to spread northward beyond the Colorado Plateau onto the fringes of the Great Basin. Whether the farmers who occupied this region during this period were Great Basin peoples who adopted agriculture and a Puebloan lifestyle, or alternatively whether the ancestral Basin hunter-gatherers temporarily were pushed aside by Puebloan groups moving north, is not known. Many archaeologists favor the hypothesis that the Valley of Fire was occupied by Puebloan peoples who pushed northward.

It is clear, however, that after about A.D. 1200–1300, a time of great drought, this portion of southern Nevada was occupied by the Numic ancestors of the Southern Paiute. More to the point is the fact that Atlatl Rock contains petroglyphs that are fully characteristic of the prehistoric and ethnographic cultures of the Great Basin, but it also contains some motifs that are more typical of Puebloan rock art sites and presumably date to the period when farming was practiced here. These include birds drawn in profile, plants, outlined stars, and phallic stick-figure human motifs with bent arms and legs. Although we do not know whether the Puebloan style rock art was also made by shamans to depict vision quests, like the Numic petroglyphs, it is clear that Atlatl Rock was a place where cultures intermixed, and a place that all groups signaled as sacred. Regardless of a specific culture responsible for any given petroglyph, the Atlatl Rock petroglyphs exhibit a strong Numic influence. Indeed, the similarity in subject matter between Atlatl Rock and other Numic sites is so strong it precludes the possibility of coincidence, suggesting that we may use Numic ethnography to speculate about the meaning of these petroglyphs more generally.

Like many other rock art sites, including those of the Numic speakers, the Atlatl Rock petroglyphs are dominated by the entoptic patterns common during the initial stage of a vision. A smaller number of other petroglyphs, including bighorn sheep, human figures, footprints, and a horned lizard (the "horned toad"), are also present. Particularly notable at the site — and the reason for its name — is a realistically engraved throwing board or atlatl (photo 31). Indeed, this is one of the best prehistoric renderings of an atlatl in North America; more typically, atlatl petroglyphs appear to have been schematized as a small circle bisected by a long line. As has been noted previously, atlatls were used as hunting implements to launch small dartlike spears prior to the appearance of the bow and arrow, which occurred at approximately A.D. 500. Although the complete replacement of the atlatl by the bow and arrow may have taken a few centuries to effect, as a general rule of thumb we can consider atlatl motifs to be greater than 1,500 years in age. We can infer, then, that at least some of the petroglyphs at Atlatl Rock are 1,500 years old or more, although it is also likely that some of the engravings at this site were made both earlier and more recently.

The presence of what seem to be relatively large numbers of atlatls and, subsequently, bows and arrows in rock art, along with depictions of "game," "hunters shooting game," and so on, led many early researchers to hypothesize that the art largely concerned hunting, in general, and perhaps "hunting magic" specifically. We now know that this hypothesis was incorrect: Such hunting-themed art is actually relatively rare overall, thus indicating that the hunting-

Photo 31: A high petroglyph panel at Atlatl Rock in Nevada's Valley of Fire. Note the realistically rendered atlatl engraving at the top and the long dart or throwing spear below it. Bow-and-arrow technology replaced atlatls about 1,500 years ago, indicating that these petroglyphs predate A.D. 500. (Scale: atlatl is about 3 feet long.)

magic hypothesis would only explain a small portion of the art anyway. The ethnographic record provides us with an alternative understanding of the origin and meaning of the art. In Nevada, for example, hunting-themed art constitutes less than 10 percent of the known petroglyphs. We obtain an inflated perspective of the importance of hunting-theme motifs because our attention is drawn to such identifiable designs much more than to the considerably more common, but interpretively more elusive, entoptic patterns. Hunting weapons like the atlatl, and hunting metaphors like killing a bighorn sheep (see site 1), were commonly employed in shamanistic rituals and beliefs because, in the Far West, shamanism was largely an adult male activity and the creators of this art were hunters. In the historical period, shamans were said to sometimes "receive" through visions bows and arrows as ceremonial objects, and would use them as ritual paraphernalia during curing ceremonies. Similarly, they would occasionally receive the supernatural power in their trances to cure arrow wounds or, subsequently, gunshot wounds as signaled by visions of weaponry and warfare. At this point in time, our best hypothesis is that similar beliefs were maintained by the earlier creators of the atlatl motifs; thus, it is likely that this motif represents the work of an Atlatl Shaman, an individual who specialized in the treatment of spear wounds.

Slightly below the atlatl and in the center of the same panel is another important motif: an upright, bipedal bighorn sheep that is a conflation of a human and a bighorn, displaying the fact that shamans were transmogrified, in the supernatural world, into their spirit helpers, in this case bighorn sheep (photo 32). The association between bighorns and the shaman's world of trance is further emphasized by another engraving slightly to the right of the human-sheep conflation. This is a relatively large bighorn with a wavy line emanating from its snout. Upon going into a trance, many shamans would bleed from the nose and/or mouth, apparently due to hyperventilation and the rupturing of nasal membranes (see site 29). Sheep with emanations from the snout can be inferred to represent bighorns bleeding from the nose—that is, Bighorn Sheep Shamans in a trance, rather than "real" sheep.

Photo 32: This is the central portion of the petroglyph panel shown in photo 31. Note the upright bipedal bighorn at center-left, representing a shaman transformed into a bighorn. Note also the wavy line emanating from the mouth of the large bighorn in the center of the frame; this probably indicates bleeding from the nose and/or mouth—a common result of hyperventilation and rupture of the sinus membranes while in a trance. (Scale: central bighorn is about 8 inches long.)

Map 16: Atlatl Rock and Mouse Tank.

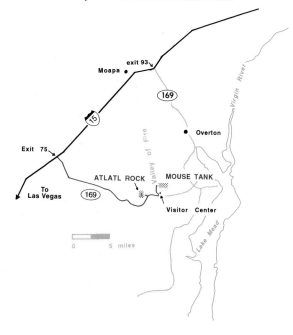

▰▰▰ **Visiting Site 24:** Visit Atlatl Rock in conjunction with Mouse
Tank petroglyphs (site 25).

Atlatl Rock is within Valley of Fire State Park. It can be reached
by taking Interstate 15 north from Las Vegas for approximately 34
miles (map 16). Turn right (east) at exit 75, which is Nevada Highway
169. The park is about 15 miles from I-15 down this all-weather dirt
road. Signs will direct you to the rock art site, which is approximately
2 miles beyond the entrance. Note that while the main panel can be
examined from a viewing platform at the top of a large set of stairs,
additional petroglyph panels are present around the rock and on other
rock outcrops nearby.

For information call the Valley of Fire State Park Visitor Center
near Overton (see appendix for phone number). There is a small fee
for entering the park for day use or camping.

SITE 25
Mouse Tank Petroglyphs
Valley of Fire State Park, NV

Sheep and Dancers

The second site at the Valley of Fire State Park is the Mouse
Tank Petroglyph Trail. It consists of an impressive series of panels
located along a short trail to a deep depression in the rocks that
collects and stores water seasonally. Like Atlatl Rock, Mouse Tank
falls within a region that was occupied by Puebloan farmers during the period from approximately A.D. 1 to 1200 and, like this first
site, Mouse Tank contains many Puebloan-style petroglyphs.
Indeed, Mouse Tank appears to be more purely Puebloan in age
than Atlatl Rock, which includes motifs of pre-Puebloan, Puebloan,
and post-Puebloan or Numic ages. The Mouse Tank Trail affords
a rare opportunity to view a kind of site in southern Nevada that
typically would require traveling to Arizona or southern Utah to see.

The interpretation of Puebloan rock art is, at this point, still prob-
lematic. Southwestern farming cultures suffered great disruptions

at approximately A.D. 1200, a time of significant drought resulting in the abandonment of many sites and, in certain cases, entire regions, including the Valley of Fire. No one is certain of the implications these disruptions may have had for religion and belief, although it is likely that there were major changes in these facets of culture. Thus, we cannot easily use recent ethnography from the Hopi, Zuni, or the Rio Grande Pueblos to interpret the pre-A.D. 1200 Puebloan rock art.

That said, there are nonetheless a few points that can be made to give us some understanding of the Puebloan petroglyphs. The first concerns their prehistoric link with the hunter-gatherer art of California and the Great Basin. In both cases these two art traditions derive from a shared older archaic substrate of religion and belief. We have every reason to assume that this substrate was shamanistic; in fact, there continue to be shamanistic elements in Puebloan religions, thereby affirming their shamanistic origins. Furthermore, we know that one of the characteristics of these archaic shamanistic practices was the making of rock art. Some aspects of Puebloan rock art, therefore, are likely to be shamanistic in origin and intent.

This last supposition is well supported by the art, especially with reference to the neuropsychological model of motif forms that derive from altered states of consciousness, discussed in the introduction. At Mouse Tank we see many, if not all, of the entoptic patterns that are common percepts in the first stage of a trance. They include zigzags, meandering and parallel lines, patterns of dots and circles, spirals and concentric circles, and nested curves, as well as more complex combinations of these geometric forms. The presence of this suite of entoptic motif types makes it unlikely that they derive from sources other than altered-states mental imagery. However, as I have emphasized a number of times, this inferred fact concerning the origin of the art tells us little about the cultural meaning of the motifs. For example, although the recent ethnography provides us with an understanding of the intended meaning of spirals and concentric circles in the Far West — whirlwinds, the concentrators of supernatural power, and the shaman's ability to fly (see site 8) — it is simply unknown whether or not the whirlwind interpretation of this entoptic pattern might also apply to Mouse Tank's Puebloan rock art.

The situation becomes more complex when the representational figures at the site are examined. You will see bighorn sheep, but also deer and, especially, human motifs on the panels at Mouse Tank. Although these kinds of motifs are similarly present in the hunter-gatherer art of the Far West, the Puebloan examples tend to be more systematically stylized and formulaic than seen in California and the Great Basin. This is particularly true of the human figures: the idiosyncratic "patterned-body anthropomorphs" that typify Numic rock art, for example—each of which displays a unique internal body design composed of entoptic patterns—are replaced by smaller, solid-body humans, many of which are identical or near-identical in form in the Puebloan art. The near carbon-copy duplication of particular motifs, then, is a trend in Puebloan rock art, contrasting with the more idiosyncratic renderings of a set of motifs in the hunter-gatherer art.

There is no better illustration of this than the panel of human and animal petroglyphs and entoptic designs at the site (photo 33).

Photo 33: An ensemble of four human-figure petroglyphs (top left) and a variety of other motifs, including bighorns and geometrics, at Mouse Tank. Unlike most of the sites in this guide, Mouse Tank appears to have been made by Puebloan farming peoples rather than by hunter-gatherers. The petroglyphs at this site, accordingly, are more typical of rock art in Arizona and Utah than in California and Nevada. (Scale: human figures are about 8 inches high.)

Particularly notable is an ensemble of four human figures, holding hands and standing in a line. The two figures on the right are shown with thin, straight torsos and waists; those on the left are blocky, rounded, and almost obese. What is most interesting in this case is that precisely the same motif ensemble is present elsewhere on the site: four humans holding hands, with two thin figures on the right and two blocky figures on the left.

The implications of this repetition of complex images are straightforward, the first being the existence of formalized and organized religious beliefs and practices, removed from the more idiosyncratic nature of the individualized shamanistic practices of hunter-gatherer groups. At the Mouse Tank petroglyphs, as well as those at Atlatl Rock, we see an expression of what are presumably formal religious cults and rites such as those still practiced by Pueblo groups today, rather than simply the practices of a lone shaman or a few ritual initiates. Second, it is then unlikely that the figurative images are all necessarily depictions of spirit helpers, and other spirits and beings, in the supernatural world. Instead, they may represent deities, mythic actors, ritual participants, and so on, with the potential meaning of this Puebloan art a reflection of the more formal and structured nature of Puebloan religion relative to that of the hunter-gatherer.

▲▲▲ **Visiting Site 25:** Visit Mouse Tank in conjunction with Atlatl Rock (site 24).

From Atlatl Rock (see map 16) head east for 2 miles to a marked turnoff on your left, leading you to a new visitor center and Mouse Tank. The parking lot to Mouse Tank is about 2.3 miles beyond the turnoff, and is well marked. The trail toward the tank is roughly 0.5 mile round-trip down an easy path, but you will see the first petroglyphs on your left about 150 yards from the car park. Most of the panels are located on your left as you walk down to the tank.

Ignore the sign at the trailhead, which suggests interpretations for some of the motifs that are fanciful but dubious. Also ignore the caption in the visitor center, which states that "[n]either science nor history will reveal the meaning of the petroglyphs"; it was written before either science or history had been examined to learn about the art. For information, call the Valley of Fire State Park Visitor Center (see appendix). A small fee for entering the park is charged for day use or camping.

SITE 26
Willow Spring Rock Art
Red Rock Canyon, NV

Touching the Spirit World

Probably no rock art motif is more universal than the hand-print. Whether the site in question is in North America, southern Africa, Australia, or western Europe, and whether painted or engraved, handprints are seemingly everywhere present in small but significant numbers. Yet the apparent commonality of this motif contrasts with its evocativeness; the handprint speaks to the inherent humanness of rock art in ways that bighorn sheep, entoptic patterns, and strangely rendered human figures never approach.

The shared humanness that the handprints evoke, I suspect, results not so much from the self-evident characteristics of these motifs—their obvious human and inherently personal origin—as from the fact they speak to our strongly tactile natures. As any archaeologist working with the management of rock art sites knows too well, the urge to touch is universal and compelling; manage-ment strategies for our fragile sites literally include, as a major component, strategies to subvert this so-human inclination. The evocativeness of the rock art handprint, then, is not so much based on the physical fact that we all have hands; rather, it is due to a recognition that the visitor shares the same subconscious urge to touch the rock face that motivated the prehistoric artists.

Willow Spring, the first of two sites in the Red Rock National Conservation Area outside Las Vegas, has a particularly well-preserved panel of five red-painted pictographs (photo 34). As is fitting for a rock art motif that is so universal, we are uncertain as to which Native American cultural group was responsible for this panel and the other rock art at this site, as well as the age of this art. As implied with reference to the other southern Nevada sites (see sites 23, 24, and 25), this region experienced considerable cul-tural flux at different times in the past, and historically sat near the boundary between the Yuman speakers along the Colorado River and the Numic speakers (Southern Paiute) of the Great Basin

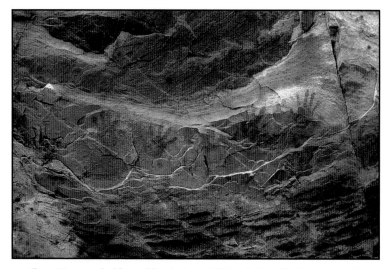

Photo 34: A panel of five red handprints at Willow Spring. The frequency with which handprints appear at sites reflects the belief that rock faces were thought of as permeable boundaries between the supernatural and natural worlds. By touching the rock, shamans in essence touched the sacred realm. (Scale: handprints are about 6 inches.)

proper. Either of these two groups are likely candidates as the creators of this art. While each group suggests different implications for the finer points of interpretation of the different motifs present here, at the level of origin commonalities still existed between Yuman and Numic rock art, thereby informing a general interpretation of the panels.

The five red handprints, for example, apparently reflect the belief that the rock face was a permeable boundary between the natural and supernatural worlds. It was the door the shaman entered to visit the spirits, regardless of whether these spirits were the gods and actors of mythic times, as among the Yuman groups, or the spirit helpers, ghosts, and other supernatural beings that occupied a sacred world unconnected to mythology, as with the Numic. Because the rock art site was a sacred place imbued with supernatural power and inhabited by spirits, the rock face was more than simply a convenient boulder for the shaman-artist, and more than a neutral backdrop or canvas for his art. Instead, it maintained a material spirituality much like the symbols themselves (see site 30). By touching the rock face, and by then recording this event in the same manner that shamans recorded the visions that they saw "inside" the rock art site when they traveled into the spirit world, the shaman at once absorbed some of the power contained within the rock and affirmed his intimate connection with the supernatural.

In addition to the panel of red handprints, Willow Spring also contains three panels of petroglyphs and an additional red pictograph, located across a wash and slightly upstream from the handprints (photo 35). These additional motifs are all entoptic patterns, the geometric images perceived during the first stage of an altered state of consciousness, thereby demonstrating the use of this site as a vision quest locale. The images are dominated by meandering lines, a rectangular grid and "rake" motifs, and an engraved web-like grid, alone on a panel under a shallow overhang.

As is particularly true for many Numic rock art sites, a village was once located at the spring at this locale, as is evident in the dark, organically enriched midden soil in the picnic area around the spring head. Again, this emphasizes the connection between

Photo 35: This netlike design in a petroglyph at Willow Springs shows one of the many variations on entoptic patterns. (Scale: motif is about 18 inches wide.)

vision quest sites and permanent water, based on the belief that springs were inhabited by supernatural spirits.

◤◣◢◥ **Visiting Site 26:** Visit Willow Spring in conjunction with Red Spring (site 27).

Willow Spring is located in the Red Rock National Conservation Area, which is about 14 miles west of downtown Las Vegas. To reach the site, head west on West Charleston Road from I-15 in the middle of Las Vegas (map 17). You will encounter a paved turnoff on your right about 16 miles from I-15, which leads to the BLM Visitor Center and Red Rock Scenic Loop Drive. Take this one-way scenic loop for about 7.6 miles, which will bring you to the Willow Spring Picnic Area turnoff, also on your right. Proceed about 0.5 miles up this road to the first picnic table in the Willow Spring area, which will be on your right. The panel of handprints is located on a cliff face roughly 25 yards beyond this picnic table; a small sign at the base of the panel will help orient you. Although it probably goes without saying, *please* do not touch the handprints; they are still well preserved only because most people never realize the prints are there, and thus they have not

Map 17: Willow Spring and Red Spring.

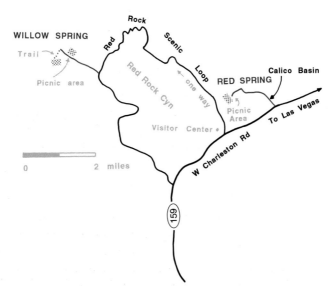

been subjected to the perhaps inadvertent but real chemical and physical degradation that results from visitors placing their hands on top of the painted motifs.

The petroglyph panels and additional pictograph are located on the large cliff face across the wash from the picnic area. This cliff face is covered by vegetation, and finding the panels requires a little searching among the bushes. The easiest way to locate them is to proceed to the far end of the picnic and parking area, about 0.1 miles beyond the first picnic table. A signed trailhead begins near the last picnic table, which is on the wash side of the road. Follow this trail across the wash to the opposite cliff face, then search along the base of the cliff toward your left. The panels are dispersed along a roughly 50-yard stretch of the cliff face. For more information, contact the Red Rock Canyon National Recreation Area (see appendix for phone numbers).

SITE 27
Red Spring Petroglyphs
Red Rock Canyon, NV

Material Culture and Power Objects

The second rock art site in the Red Rock National Conservation Area is Red Spring. This is a moderate-size petroglyph site spread across a boulder scree slope and series of cliff faces, adjacent to a permanent spring and the remains of a village. As at nearby Willow Spring (site 26), the rock art creators of Red Spring are not known, and for the same reasons: At different times, Numic, Yuman, and even Puebloan peoples lived in southern Nevada and, until the art at this site is chronometrically dated, it is impossible to determine with any certainty which group was responsible for its creation. That said, it is nonetheless clear that the petroglyphs at Red Spring bear little resemblance to the Puebloan rock art of the Valley of Fire region (see sites 24 and 25); indeed, based on stylistic impressions, the site looks most like the Colorado River sites (see site 23), suggesting a Yuman origin. Yet even a solely Yuman origin seems unlikely, simply because we know the Numic occupied this area historically and Numic shamans commonly conducted vision quests at springs and engraved petroglyphs as a result. Given the current lack of hard empirical evidence, our best hypothesis is that the rock art at Red Spring resulted from Yuman and Numic traditions conducted at the same place at different times. This again points to the Native American sense of sacredness, which was imbued in specific places on the landscape and often transcended cultural boundaries (see site 24). Moreover, these cultural and temporal inferences are supported by the revarnishing that is evident in the petroglyphs, which ranges from nonexistent to moderately heavy. In a general way, this suggests that some of the petroglyphs at Red Spring may be relatively recent while others may easily be a few thousand years in age.

As at most sites, the motifs at Red Spring are dominated by entoptic patterns, including meandering lines, sinuous snakelike patterns, amorphous grids, and patterns of dots. Vertical lines running through columns of circles or triangles are also common. At least one handprint and one zoomorph—a heavily revarnished, squat-bodied, tailed creature, perhaps a desert tortoise—are also present. An unusual petroglyph, located on a cliff face above the scree slope, is particularly interesting. Representing an integration of different entoptic designs, it is nonetheless clearly a depiction of a fringed blanket, robe, or bag (photo 36). That this relatively unique motif is both authentic and relatively ancient, perhaps on the order of 1,000 years in age, is guaranteed by the moderately heavy revarnishing it exhibits.

The presence of this depiction of Native American "material culture" immediately raises a series of questions, especially in light of the general interpretation of rock art as largely originating in visionary images that I have presented throughout this guide. The first question concerns origin, i.e., whether or not hunter-gatherers such as the Numic or the partly horticultural Yumans made and/or used blankets, robes, or bags in any way similar to the one shown in this petroglyph. Simply put, the answer is yes, once it is recognized that such a blanket need not have been woven, but instead

Photo 36: This petroglyph panel at Red Spring evokes the image of a blanket. During a shaman's trance he "received" (that is, dreamed about and later constructed or acquired) material-cultural items such as talismans, wands, and ritual costumes, which he then used during subsequent ceremonies. This blanketlike motif likely represents one such "power object" a shaman saw during his trance. (Scale: blanket image is about 2.5 feet long.)

may have been a fringed animal hide with a painted design. Painted hide objects, including quivers, shirts, shields, and, among the Numic-speaking Northern Shoshone, tipi covers, were common features in hunter-gatherer material culture. Still, this need not imply that the blanket in question was necessarily hide; Puebloan groups in southern Nevada and other portions of the Southwest certainly practiced weaving. Inasmuch as we have archaeological examples of Pueblo pottery that were traded as far west as the Los Angeles Basin, it is clear that much more interchange occurred between different Native American groups in western North America than we otherwise might assume. The Yuman speakers along the Colorado River, such as the Mojave, were particularly instrumental in conducting trade between Californian groups and the Pueblos of the Southwest. Among other items, they are known, historically at least, to have transported blankets from the Southwest to the Pacific Coast, returning with ocean products such as shell beads. The decorated blanket/bag motif at Red Spring, then, may represent a product of a hunter-gatherer culture or, if in fact it was woven, a trade item obtained by hunter-gatherers from Puebloan groups.

If we assume for a moment that it was not woven but instead represents a painted hide, a second question concerning origin occurs: Where did the design pattern for this article originate? Clearly, the motif incorporates designs that resemble the entoptic patterns of a shaman's trance, and it requires no inferential stretch to suggest that the blanket is in fact an entoptic pattern that was construed as a blanket during the second stage of an altered state. Given this possibility, two recorded ethnographic comments are enlightening here. Though they pertain in both cases to basketry designs, they indicate that designs such as this were widely recognized as similar to the entoptic patterns seen during trances, and that, most importantly, all graphic designs were believed to originate in "dreams"—that is, in the supernatural world, the world of spirit patterns. From the Native American perspective, then, any graphic design was de facto a spirit pattern, regardless of whether placed on a mundane object or a sacred artifact.

The final question is one not so much of origin as of use: What purpose would a blanket or bag have played in religious practices, and how does the presence of such a motif fit with the larger interpretation of rock art as the product of vision quests? Two factors are at play here. The first is simply the fact that shamanistic visions, and thus shamanistic rituals and religion, were at one level highly idiosyncratic precisely because they involved a shaman's direct participation with the supernatural. This obviously could vary in specific details from one altered state of consciousness to another. More to the point is the fact that a shaman "received" his ritual talismans—that is, objects imbued with power that he used in curing and other ceremonies—while in the supernatural world, meaning that he had visions of them. Following their use, and after their power was "used up," the shaman returned to the supernatural to renew their potency. A series of items were common "power objects" for the shamans, such as rock crystals, feathers and bird down, crooked staffs, and snake and animal parts such as claws, fangs, beaks, and wings. Yet some aspects of the shaman's kit of power objects were less common or even idiosyncratic. These included feathered headdresses, strings of beads, deer hoof rattles, bows and arrows, stone and wooden knives and daggers, "scepters" and wands, and a miscellany of other items, each of which was kept secreted away in an animal hide bag or wrap, usually made of badger or weasel skin. Not surprisingly, these so-called medicine bags are relatively common in rock art, particularly in the petroglyphs of the Coso Range.

The presence of a petroglyph possibly portraying a blanket at Red Spring, then, fits a larger pattern in which material-culture items of various sorts were perceived by a shaman while in a trance, and subsequently used as power objects in his ritual activities. Although blankets are relatively rare in rock art, they are not entirely unheard of. If this motif is actually intended to portray a bag, then a significant number of parallel examples from other sites in the Far West could also be cited. The important point, however, is that there were few if any a priori constraints to what could become a shaman's power object, and the blanket motif at Red Spring was most likely intended to portray one.

ΛΛΛ Visiting Site 27: Visit in conjunction with Site 26.

The Red Spring petroglyphs are also located in Red Rock National Conservation Area, about 14 miles outside of Las Vegas. To get to the site, follow the directions to Willow Spring (site 26) by taking West Charleston Road out of Las Vegas (map 17). You will encounter a marked turnoff to Calico Basin on your right about 14 miles west of I-15, or about 2 miles before the turnoff to the Red Rock Canyon Scenic Loop. Follow this road through a rural residential area for about 2 miles, at which point you will encounter a T-shaped intersection. The turn to Red Spring, which is marked, is the left fork of this intersection. This road will lead you to the Red Spring picnic grounds, which consist of three parking areas and associated tables.

The petroglyphs are located on the scree slope and cliff faces behind the tables. A moderate number of motifs are present at this site, although the panels are relatively widely dispersed and some searching and climbing is required to see all of the art. For more information, contact Red Rock Canyon National Recreation Area (see appendix).

MAP 84
c.2

Southern San Joaquin Valley and South-Central Coast Region

SITE 28
Hospital Rock Pictographs
Sequoia National Park, CA

The Drowning Shaman

Hospital Rock is located in Western Mono territory, within the southern Sierra Nevada. The Western Mono are Numic speakers, linguistically related to the Owens Valley Paiute. At some point in the not too distant past, however, they migrated west across the crest of the Sierras, to live alongside the Yokuts. As a result of this contact, the Western Mono adopted many of the cultural traditions of their western neighbors, including the painting of pictographs at sites, following the pattern of south-central California groups rather than continuing with the petroglyph tradition of their Numic kin to the east.

As is typical of rock art sites in south-central California, Hospital Rock served to record the visions of male shamans, in this case two brothers who resided in the adjacent village during the historical period. That the site can be linked to two historical shamans indicates that at least some of the pictographs are less than about 200 years in age. The site also provides a good example of how place names, historical tales, the setting, and the art itself all serve to encode and express the shamanistic nature of south-central California rock art sites.

According to the local ethnographer and historian Frank Latta, the aboriginal name of Hospital Rock was *pahdin*, which translates into English as "place to drown." Hospital Rock, however, is some distance from the river, indicating that pahdin was intended as a metaphoric rather than literal descriptive name. In this case, "place

151

to drown" alludes to drowning or going underwater as a metaphor for the shaman's entry into the supernatural. The logic of such a metaphor is based on physiological analogies between going underwater and entering a trance. In both instances, an individual may experience blurred vision, impaired hearing, and body movements that are at once weightless but also sluggish and difficult (see site 34). The place name pahdin, then, can be understood as signifying the "place where the shaman entered a trance."

An historical tale linked to the site recounts an incident in which the two shaman brothers encountered, and overcame, a grizzly bear in the rock at Hospital Rock. Although the tale is often taken at face value—as a true event—it is most likely a reference to the shaman brothers' vision quests. Throughout the Far West these were believed to occur when the shaman "entered the rock," generally through a crack or tunnel that would open at his command (see site 31). His visionary experience then began when he crossed a supernatural rattlesnake and grizzly that guarded the entrance to the sacred (see site 29). A discussion of two shamans' involvement with a grizzly inside the rock at a rock art site, then, is almost certainly a reference to supernatural events said to occur during the shamans' trances: the "test" of their worthiness for supernatural power by battling dangerous spirits at the outset of their visionary experiences.

The setting of the paintings at Hospital Rock further reflects the connection between them and the supernatural. As noted above, the doors of the supernatural were believed to be cracks in the rock that opened for the shaman when he ingested his preferred hallucinogen, native tobacco, and entered an altered state of consciousness. Motifs at many sites, accordingly, are portrayed as coming out of cracks in the rock, as if emerging from the supernatural (see sites 18 and 19). At Hospital Rock, the painted panel itself is placed on one side of a massive split in the boulder, suggesting that this crack was the shamans' entrance into the sacred.

The motifs painted at the site are all red, as is typical of many pictograph sites in the Far West (photo 37). Most of the motifs are entoptic patterns, the geometric designs experienced during the initial stage of the shaman's trance. At least one rattlesnake—with grizzly, a guardian of the supernatural—is also present on the panel, along with some schematic anthropomorphic figures, which could represent either humans or bears.

⋀⋀⋀ **Visiting Site 28:** Hospital Rock is located within Sequoia National Park, immediately adjacent to California Highway 198 between Three Rivers and Giant Forest Village (see map 18). The well-marked site, complete with paved parking, interpretive display, and restrooms across the road, is approximately six miles east of the Ash Mountain park entrance. Although the highway into the park is paved and open all year, during the winter it may be snow covered, requiring chains or four-wheel drive. There is an entrance fee to the park. For more information, contact Sequoia and Kings Canyon National Parks (see appendix).

Map 18: Hospital Rock.

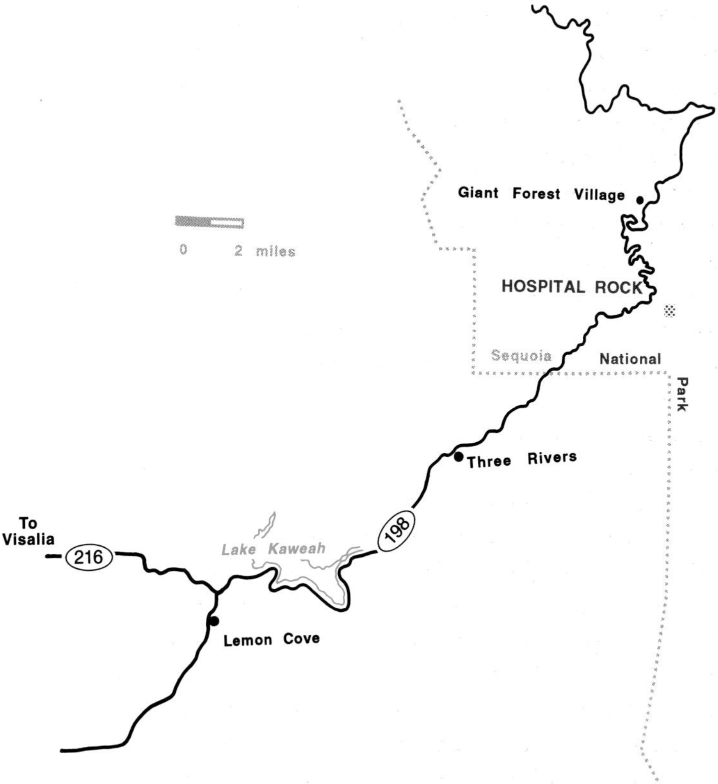

Photo 37: A red pictograph panel at Hospital Rock. Ethnographic accounts indicate that this was the rock art site of two brother shamans who probably lived here during the latter half of the nineteenth century. Most of the motifs at this site are entoptic patterns, the precise meaning of which is unknown, but there are several recognizable designs on this panel such as the zigzag (a rattlesnake) at the upper left and the human figures. (Scale: zigzag rattlesnake is about 18 inches long.)

MAP 84
G-2

SITE 29
Tulare Painted Rock Pictographs
Tule River Indian Reservation, CA

Guardians of the Supernatural

Tulare Painted Rock is located in Yokuts territory, which stretched from the southern end of the San Joaquin Valley to the Sacramento Delta, and included both the valley floor and the foothills of the Sierra Nevada. This was a particularly bountiful region, covered by sloughs and swamps in the valley bottom, which were filled with fish, water fowl, and game. Rolling hills dotted with oak forests graced the higher elevations. This site is one of the most spectacular of the Yokuts sites, if not California sites generally. Rendered in red, white, yellow, and black, it exhibits a complex of images that serve as a textbook example of how south-central California beliefs about the supernatural world were portrayed by the shaman. A lizard, for example, has been painted at the entrance to this large boulder cave. Lizards move in and out of cracks in the rocks which, as noted previously, were believed to be entrances into the supernatural. Lizards were thus considered messengers between the sacred and mundane worlds, with their common depiction at rock art sites signaling that such served as doorways to the supernatural.

A zigzag-rattlesnake motif is located immediately inside the drip line of the cave, just beyond the lizard. In south-central California, rattlesnakes, paired with the grizzly, were believed to guard the supernatural. As you enter the cave, your eyes will be drawn to the back wall where a series of large grizzlies are painted. The painting on the extreme right of the back panel (photo 38), closest to the

Photo 38: A painted panel on the back wall at Tulare is dominated by this large red and white grizzly bear (standing, with its arms splayed). Careful examination of this art reveals facial qualities characteristic of grizzlies. Shamans in this region believed grizzlies and rattlesnakes were paired as dual guardians of the supernatural; to enter the sacred realm, the shamans had to pass them. A rattlesnake motif is also present at this site, on the ceiling panel at the drip line (top right). (Scale: bear motif is about 4 feet high.)

rattlesnake, is particularly notable. Although faded, this large figure is portrayed with the characteristic facial exudations of the grizzly, resulting in dark streaks down their cheeks. This characteristic sometimes resulted in references to grizzlies as "old pitch on the face," and was believed to connect the grizzly with the shaman who sometimes bled from the nose or mouth during his trances (see site 24). The shaman-artist, then, chose to illustrate the two guardians of the spirit world, the rattlesnake and grizzly, which he passed when he entered the sacred realm.

There are a number of grizzly and other paintings on the back wall, including a series of three human figures with stretched and elongated heads, depicting one of the bodily hallucinations common to shamans in the trance state. A small panel of yellow motifs, farther within the cave along the back wall, illustrates human fig-

Photo 39: A centipede pictograph on the ceiling panel at Tulare. In aboriginal southern Sierra Nevada, centipedes were erroneously believed to be deadly poisonous and therefore potent spirit helpers for shamans. (Scale: centipede is about 4 feet long.)

156

ures with bifurcated "exploding" heads, another rendering of this same bodily hallucination. But the most spectacular panel is the ceiling of the cave itself. This includes a series of large human figures, a centipede (photo 39), a beaver, a frog, and a large yellow animal identified as *soksouh*, a "dangerous" supernatural spirit. Numerous entoptic patterns are also present, adjacent to and sometimes superimposed over these other images.

Although not in fact poisonous, centipedes were considered to be malevolent in south-central California and were therefore particularly powerful spirit helpers. Similarly, the beaver was a specialized spirit helper for the Ohowish Shaman, a shaman whose powers were connected with water and finding lost objects. The beaver and the frog motifs also allude to the metaphor of going underwater for the shaman's trance (see site 28), in this case a particularly apt supernatural metaphor given the pool of water that is immediately adjacent to the site.

This site was called *Uchiyingetau*, or "markings." It appears to have been painted by a known historical shaman, and thus was created between 100 and 150 years ago. It is a well-known site and has been described in the professional and popular literature a number of times since 1902. Like many such well-known sites, popular discussions of it have contributed unfortunate and fanciful myths. One of these is that the large grizzly on the back panel is Bigfoot; another is that the soksouh figure is "coyote eating the sun." Neither of these interpretations is supported by the ethnography specific to the site, nor that of Yokuts people in general.

▰▰▰ **Visiting Site 29:** This site is located on the Tule River Indian Reservation, east of Porterville, California (map 19). It can be reached by taking California Highway 190 east out of Porterville toward Lake Success. Turn right (south) on County Road J42 about 5.9 miles east of the junction with California Highway 65. Stay on J42 as it winds around, eventually heading east. You will enter the Tule River Reservation about 9.5 miles beyond California 190. Tulare Painted Rock is another 5.1 miles farther along. As you approach the site, you will pass a sawmill on your left (north); shortly thereafter, you will encounter a sign to the site and a dirt parking lot on your right (south). The site is to the immediate southwest of the parking area,

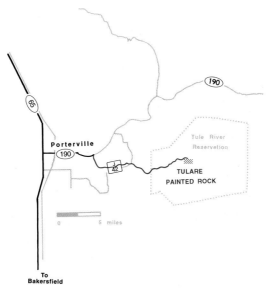

Map 19: Tulare Painted Rock.

on the south side of a very large boulder. For further information, contact the Tule River Indian Reservation headquarters (see appendix).

MAP S3
E-11

SITE 30
Rocky Hill Pictographs
Exeter, CA

Continuing Cultural Traditions

No California site better illustrates the importance of rock art to continuing Native American beliefs and traditions than Rocky Hill, outside of Exeter, Tulare County. In part, this results from the on-going involvement of local Native Americans with this locality; but it is also due to the quality and importance of this major expression of rock art. Unlike most south-central California rock art sites, which consist of one or at most a few painted panels localized around a single rock shelter, Rocky Hill contains numerous polychrome painted shelters, dispersed across a mazelike, boulder-strewn hillside. This concentration of different panels is similar in some respects to Carrizo Painted Rock (see site 32), which also stands out as significantly larger than the typical pictograph site. In both cases, the quantity of paintings suggests that these localities were used by a number of different shamans, probably for many hundreds, if not thousands, of years. Since rock art marks sacred spots on the landscape, we can only infer that Rocky Hill and Carrizo Painted Rock were widely recognized as particularly sacred locations.

Rocky Hill falls within the territory of the Wukchumni, a Yokuts-speaking group that lived along the banks of the nearby Kaweah River, and continue to live in Tulare County today. The remnants of a large Wukchumni village are located immediately adjacent to the hillside containing the pictograph panels. The site area, however, was acquired by Euro-Americans during the historical period. While the local rancher is to be greatly commended for having maintained the pictograph panels in pristine condition, Euro-American occupation and use of the property nonetheless limited Native American access to it. Recently the Archaeological Conservancy, a nonprofit organization dedicated to the preservation of sites, acquired a portion of Rocky Hill containing some, but not all, of the major rock art panels. In their efforts to manage the site, the Conservancy has sought the input of the Wukchumni. This process has revealed two heretofore unexpected facts. First, it has become apparent that the cultural importance of Rocky Hill was maintained throughout the approximately 100 years during which it was out of Wukchumni control, and that the Wukchumni, necessarily surreptitiously, used the locality for ritual purposes during that period. Second—and completely disproving many archaeologists' beliefs that historical, let alone modern, Native Americans knew nothing about rock art—the Wukchumni maintain considerable traditional knowledge about the panels. This revelation has been particularly gratifying to me, as an archaeologist, because the modern Wukchumni have provided direct confirmation of the interpretations of south-central California rock art that I developed using the older ethnohistorical literature. Indeed, Rocky Hill and the paintings it contains continue to be central to Wukchumni religious and cultural beliefs and practices.

Primary among these—and exemplary of the importance of rock art to far-western Native Americans generally—is the Wukchumni investment of a material sacredness to the painted symbols; that is, the pictograph motifs not only represent supernatural spirits symbolically, but are themselves sacred objects in a concrete sense. It is for this reason that the Wukchumni object to calling the pictographs "art," with its primary connotation of aesthetics. It is also for this reason that the Wukchumni are greatly concerned with the preservation of the pictographs. Their loss would be the loss not simply of a beautiful and scientifically important record of the Wukchumni people; it would be the loss of sacred objects that they believe are important for their spiritual survival.

The concept of investing material spirituality in a painted symbol is perhaps foreign to many Euro-Americans, due to the predominantly Judeo-Christian origin of our culture. Yet we too have similar beliefs, albeit concerning the use of words, rather than painted symbols, that parallel in some ways those of the Wukchumni. For example, though we may scoff at the fairy tale notion that the word "Abracadabra" will magically open up the locked door, we still invest magical power in the spoken word. In the act of transubstantiation, Christian ministers invoke a set of words which, in the proper ritual setting, transform the mundane objects of bread and wine into the body and blood of Christ. Ministers, judges, and ship captains also use a set pattern of words to create a physical, legal, and spiritual union between man and woman, saying, "By the power invested in me...I now pronounce you husband and wife." And by raising our right hand and repeating another set pattern of words, we guarantee that we will speak "the truth, the whole truth, and nothing but the truth," when we testify in legal contexts. In each case, the spoken word is thought to have direct material consequences, just as the Wukchumni believe that their pictographs are sacred material objects.

The pictographs at Rocky Hill, as noted above, are dispersed among a number of different panels, each of which appear to have been made by a different shaman or set of shamans from a particular family. Each panel also provides a different view of the various aspects of the supernatural world and of the shaman's interactions with the spirits. Perhaps most interestingly, Wukchumni knowledge of these different panels illustrates how the shamans' interactions with the spirit world varied, as well as how shamans' work was interpreted and used more widely in their culture by nonshamans.

A visit to the site begins at a panel near the base of the hill, where a prayer and offering is made. This is referred to as the Guardian Panel, and is painted with a bearlike motif, probably reflecting the fact that a supernatural grizzly guards the entrance to the sacred world (see site 29), and thus the entrance to the hill, which is, itself, a sacred zone. At the top of hill, a red and white T-shaped polychrome motif is said to be a "Shaman's Mark," representing the shaman as a man of spiritual power (photo 40). Similar T-shaped motifs such as these, which are common in the region, have been referred to as "pelt figures" in the archaeological literature. This modern identification tells us that these "pelts" in fact

Photo 40: The top-center pictograph in this panel at Rocky Hill was described by a modern Wukchumni informant as a "shaman's mark." Pictographs similar to this from southern Sierra sites have commonly been called "pelt figures," but the informant's comment indicates that these images represent individual shamans. (Scale: key figure is about 12 inches high.)

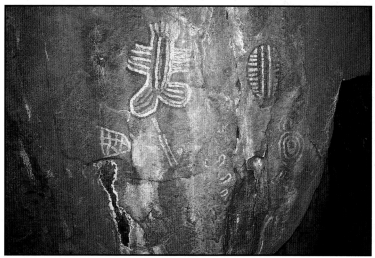

are stylized depictions of a shaman himself, probably based on a complex entoptic pattern seen in the shaman's visionary experience. Another panel is identified as Women's Panel, most likely representing the work of a female shaman (see site 21). Other panels include depictions of a Shaman's Dream (i.e., his visionary experience); Blue Heron Dancers, depicting shamans responsible for the ritual Huhuna Dance, performed as part of an annual Mourning Ceremony in a manner similar to the Shaman's Mark; evil spirits and ghosts sometimes seen in the spirit world; and so on.

▲▲▲ **Visiting Site 30:** Access to Rocky Hill is limited to group visits. These are conducted only for members of the Archaeological Conservancy, accompanied by a Wukchumni representative, about twice a year until a docent program can be organized at the site for more frequent visits. Though this may seem overly restrictive, Rocky Hill is a world-class rock art site. Since the donation required to join the Conservancy is less than the cost of most concert tickets these days, the membership donation and the 6- to 12-month wait for a site visit are small costs relative to Rocky Hill's importance. For information, reservations, and directions, contact the Archaeological Conservancy (see appendix).

SITE 31
Tomo-Kahni State Park Pictographs
Tehachapi, CA

MAP 92
G-1

Home of the Spirit Helpers

Although located in the territory of, and made by, the Kawaiisu, who were closely related to the Numic-speaking peoples of the Great Basin, the Tomo-Kahni pictograph site is more typical of the rock art sites of south-central California than the Basin: it is painted rather than pecked, and the motifs include bear and snake paintings rather than depictions of bighorns and weapons (see sites 2 and 3). This reflects the transitional status of the Kawaiisu between their linguistic and cultural cousins in the desert to the east, and their nearest neighbors in the mountains to the north and west. Although the Kawaiisu spent part of the year living on the desert, many of their cultural traditions were more akin to their south-central California neighbors than to their Numic kinfolk.

The motifs at this site are painted in red, black, white, and orange. Bear motifs are most common and have black bodies with red, or red and white, outlines (photo 41). A snake, located on a low ceiling at the back of the cave, is red with dots of white paint daubed on with a finger; as at many south-central California sites, the grizzly and rattlesnake are paired at the site as the guardians of the supernatural (see site 29). There is also a white, orange, and black human figure; black stick-figure, "bubble-eye" humans; and entoptics, including a large concentric circle, located in an alcove of the cave.

Photo 41: A bear outlined in red and white. According to at least one ethnographic account, a supernatural grizzly bear lives within the rock face of this site and guards it from unworthy intruders. This supports the belief that supernatural grizzlies and rattlesnakes were the guardians of the sacred realm. A shaman had to cross both of them to enter the supernatural world through cracks or portals in the rock. (Scale: bear figure is about 12 inches high.)

Like a number of sites in the southern Sierra, specific ethnographic information is available on this site, suggesting that it was probably created during the historical period or during the last few hundred years. The Kawaiisu generally refer to it as "Teddy Bear Cave" in reference to the bear motifs. A grizzly bear, furthermore, is said to live in a crack in the rock at the site, and is believed to emerge from the crack to frighten away visitors who approach the site without proper respect. This alludes to the belief that grizzlies, along with rattlesnakes, guard the entrances to the supernatural world. And, as noted above, bear and snake pictographs are painted, appropriately enough, at the cave.

The existing ethnographic record has also resulted in some confusion about this site. Some researchers have claimed that it represents the origin point for the Kawaiisu, based on a putative relationship between the site and a "creation myth" said to be depicted in the paintings at the site. The site, thus, is sometimes referred to as "Creation Cave." In fact, nowhere in Kawaiisu informants' statements about this site is there any mention of their origin myth. Instead, their discussions describe the origin of shamans' spirit helpers and the fact that this site, like all south-central California sites, was a home for spirit helpers. Rock art was placed in certain locales precisely because such spots were thought numinous—inhabited by spirits. The Tomo-Kahni site does not "illustrate" a creation myth, but instead reflects the intimate relationship between the shaman, his spirit helpers, and rock art sites.

Visiting Site 31: Access to Tomo-Kahni pictographs is by guided site visit only.

Tomo-Kahni (which translates as "village") State Park is located a few miles east of Tehachapi on Sand Canyon Road, north of California Highway 58. Visitation tours are held two Saturdays per month. For information and reservations, as well as directions to the meeting point, contact the Lancaster office of the California State Parks system (see appendix).

SITE 32

Carrizo Painted Rock Pictographs

San Luis Obispo County, CA

Entry into the Rock

Painted Rock is an impressive sandstone outcrop sitting on the floor of the remote Carrizo Plain, roughly midway between Bakersfield and San Luis Obispo, as a condor might fly. Lying within the historical territory of the Chumash but near their border with the San Joaquin Valley–dwelling Yokuts, this site reflects the fact that, while the Chumash are renowned as coastal and island dwellers who occupied the Santa Barbara Channel, many of their most spectacular rock art sites, and much of their population, actually fell within interior south-central California.

Carrizo Painted Rock is probably the largest Chumash pictograph site. It was also once one of the most spectacular sites in the country, if not the world (photo 42). An unfortunate episode of vandalism early in this century significantly degraded what truly had been one of our national treasures. Although this vandalism — use of the site for target practice and as a backdrop for painted graffiti—greatly detracts from it, there is still much to see, especially for the practiced and disciplined eye capable of looking beyond the vandalism to appreciate the still intact but smaller paintings, as well as the setting as a whole.

The initial impression of Painted Rock, as one approaches from the parking lot, is quite dramatic. The site comprises an isolated, massive U-shaped rock outcrop, with the vast majority of the motifs painted in a series of large panels on the interior. Numerous archaeologists and authors have noted the similarity between the outcrop's shape and a vulva. This similarity in shape is unlikely to have gone unnoticed by the creators of the art and, in fact, was probably the determining factor in shamans' uses of the site for rock painting. As noted previously, an important metaphor for the shamans' entry into the supernatural realm was sexual intercourse, due to the fact that certain hallucinogens, such as jimsonweed and

165

Photo 42: Severely vandalized decades ago, Carrizo Painted Rock was once the most spectacular rock art site in North America. This picture, taken about 1902 by Lorenzo Yates, shows the site prior to vandalism. Three seals/seal-human conflations (two horizontal, one vertical) dominate the center and left portions of the panel, while four small dancing humans above a diamond chain/rattlesnake may be seen to the right of center. (Scale: large vertical seal just left of center is about 3.5 feet high.) —Courtesy of the Santa Barbara Museum of Natural History

marijuana, act as sexual stimulants (see site 10). A shaman who experienced an altered state of consciousness, thereby accessing the sacred, did so by "entering into" the rock at his vision quest/rock art site—that is, by penetrating the symbolic vagina and symbolically performing, in a sense, a kind of ritual intercourse.

Such metaphoric action is physically manifest at Painted Rock, where walking into the site easily can be envisioned as walking into a massive vagina. Moreover, the Chumash employed another metaphor for sexual intercourse: "entering the canoe." It is not surprising that the painted panel at the apex of the interior of Painted Rock consists of four men standing in a Chumash *tomol*, or plank canoe (photo 43).

As implied above, Painted Rock has a number of large, elaborately rendered panels which are, unfortunately, in poor condition. Recent conservation efforts by an international team from the Getty Conservation Institute, however, has improved the appearance of the site dramatically from its former condition. If you look closely,

Photo 43: A pictograph of a tomol or Chumash canoe carrying dancing humans at the apex of Carrizo Painted Rock. "Getting into the canoe" was a Chumash metaphor for sexual intercourse and, in this case, appears to allude to the sexual symbolism of the site. (Scale: canoe is about 5 feet long.)

you will note fantastically rendered human figures in red, black, and/or white standing in the elbows-bent, palms-raised posture used in ceremonial dances; entoptic patterns such as circular grids and chevrons; a red snake that runs along the entire southern panel, "emerging" from and "diving" into holes and cracks in the rock face; another elaborately rendered rattlesnake, showing rattles as well as the diamondback pattern; turtles; and many other figures too numerous to mention (photo 44). And, in addition to the paintings in the interior portion of the outcrop, there are also a series of smaller panels around the outside perimeter. Although we do not know the specific age of any of the panels or motifs at the site, it is likely that all are less than roughly 2,000 or 3,000 years, with some of the paintings probably less than 500 years old.

As is clear from the number and variety of paintings at this site, Painted Rock varied in some significant ways from the typical south-central California pictograph site. In many respects it is similar to the Wukchumni Yokuts site at Rocky Hill (site 30), where a num-

Photo 44: This portion of the same panel shown in photo 42, but farther to the right, shows the current status of this panel. Though severely damaged by vandals, it illustrates the vibrant polychrome paintings that once characterized this important site. The left leg and lower torso of a standing human figure are visible, along with a series of other motifs. A polychrome turtle is just to the left of the human figure, and a second human is superimposed inside the torso of the larger figure. (Scale: left leg is about 2 feet long.)

ber of different shamans or families of shamans were responsible for creating the art. This fact speaks to the probable sacredness of both of these sites: Instead of localities owned by individual shamans, as most south-central California sites apparently were, these two spots were used for portraying the vision quests of a number of different individuals. Most likely, this resulted because they had widespread recognition as particularly powerful spots on the landscape.

Visiting Site 32: Carrizo Painted Rock was once probably the most spectacular pictograph site in North America. Historical vandalism has seriously degraded the site, but for the trained eye there is still very much to observe here. I recommend seeing Carrizo Painted Rock after visiting a few other sites, which will allow you to look beyond the vandalism to appreciate the fabulous detail that is still evident at this locale.

The site is located on the Carrizo Plain in inland San Luis Obispo County (see map 20) and is jointly managed by the Nature Conservancy and the Bureau of Land Management. You can reach the site from California Highway 58, between McKittrick and Atascadero, by turning south onto the Soda Lake Road and going through the town of California Valley. About 14.8 miles south of the highway you will encounter a dirt road on your right. This is marked with a sign to Painted Rock. The Goodwin Educational Center is 0.5 mile west down this road, with the site another 2.7 miles beyond. The site is located about 0.25 miles from the parking area down a well-maintained and easy trail.

If you are coming from the south, take Soda Lake Road heading north from California Highway 166 at Reyes Station, southwest of Maricopa. It is about a 29.5-mile stretch from the highway to the turnoff to the site. Note, however, that long sections of the southern route are not paved and are subject to flooding and wash-outs.

In addition to the potential for closed roads on the southern access route in bad weather, during certain seasons access to Painted Rock is restricted to a few guided tours per day, due to nesting birds. For information prior to your visit, contact the Goodwin Educational Center (see appendix).

Map 20: Carrizo Painted Rock.

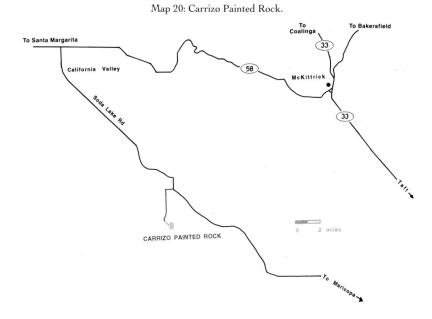

MAP 99
E-F 8

Visiting the Sites

SITE 33
Painted Cave Pictographs
Santa Barbara, CA

Dangers of the Supernatural

Probably no California site is better known than Painted Cave, a spectacular example of Chumash rock art near San Marcos Pass in the mountains immediately above Santa Barbara. This represents the historical heart of Chumash territory, which stretched from Topanga Canyon north and west to San Luis Obispo County, and inland to the western edge of the San Joaquin Valley. But it was particularly on the coast around Santa Barbara, and on the islands immediately offshore, that the largest Chumash villages developed and the most elaborate examples of their culture were found.

Painted in red, black, and white, the site contains a myriad variety of geometric patterns and highly stylized human and animal figures placed on the ceiling of a small but deeply recessed cavern. This cave has been eroded into soft sandstone; wind erosion continues to abrade the cave walls, erasing much of the art that was once present at this locale. Still, the paintings that remain warrant designating this site as one of the best extant illustrations of native Californian artistry. Although the site has not been dated, the rapidity with which this ongoing wind erosion has removed portions of the painted panels suggests that Painted Cave is no more than 1,000 years old, and probably more recent. Like other rock art sites in south-central California, Painted Cave was the work of one or a few shamans and was intended to illustrate the supernatural experiences of their vision quests.

A number of the geometric motifs at the site are sunburst or mandala-like in form. Some archaeologists have suggested that these represent depictions of the sun, solar eclipses, or other celestial phenomena, which is a hypothesis that has been widely broadcast in the popular press. While this may be so, we have no ethnographic identifications of these motifs as such, and we have considerable ethnography which supports alternative explanations of this art.

The problem with this theory, then, is not that we can prove that it is wrong, but instead that we cannot determine that it is right with any reasonable certainty.

The question of the original meaning of these sunlike paintings aside, it is clear that many if not all of these geometric motifs represent the entoptic patterns seen by shamans during their first stage of a trance, regardless of what the shamans and, subsequently, archaeologists, then interpreted the forms to be. Other pictographs at the site may be interpreted less speculatively using the existing ethnographic record, and provide us with a more confident understanding of some of the imagery of Painted Cave. For example, at least two motifs use universal far-western North American graphic conventions to represent the rattlesnake, the guardian of the supernatural and the spirit helper of the Snake Shaman. These consist in each case of a red zigzag enclosed by parallel red and white lines (photo 45). Other zigzags and diamond chains, including two

Photo 45: Painted Cave, north of Santa Barbara, is one of the more spectacular Chumash sites. Notable here is the depiction of a centipede below two zigzag-rattlesnake motifs. The centipede represents supernatural power, death, and the inherent dangers of the supernatural world. People lacking supernatural powers avoided rock art sites, or portals to the sacred realm, even when the sites lay within their villages. According to ethnographic accounts, Painted Cave was one such site. (Scale: centipede is about 12 inches long.)

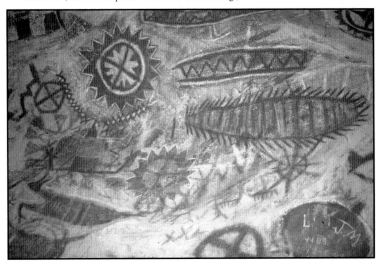

diamond-pattern chains created by rows of red Xs that are in-filled with white, also likely portray this snake spirit.

Rattlesnakes served in the capacity of supernatural guardians and particularly potent spirit helpers precisely because they were poisonous; no other snake species was afforded nearly the importance of this viper in the Far West. Since the supernatural was dangerous and its power was perilous if not used properly or if obtained by someone lacking the proper strength and skills, any spirit originating in the supernatural world was likewise potentially threatening. Following this same train of logic, dangerous animal species in the mundane were those most powerful supernaturally. Rattlesnake and grizzlies were most commonly identified as powerful and dangerous, probably due to the fact that human and animal behavioral patterns combined in such a way that these two species were commonly encountered on hunting and gathering forays. Although also dangerous, the mountain lion is nocturnal, only occasionally seen, and rarely attacks humans. It too was recognized as dangerous, and therefore supernaturally powerful, but not nearly to the same degree as the grizzly and rattler; consequently its use in rock art and ritual was much less common.

There is another important, long lozenge-shaped motif at Painted Cave, located immediately alongside one of the rattlesnakes. This is a centipede, indicated by the numerous short legs surrounding the body and the internal body segmentation. South-central California centipedes are in fact harmless, but native Californian belief maintained otherwise: The centipede bite, thought to be caused by the feet, was conceived of as incurable. Because of this belief, centipedes were strongly associated with shamans and are relatively common motifs at south-central California sites (see site 29).

The site also includes what are probably two shaman figures: long thin black-and-white checkerboards with ghostlike faces and feather headdresses—but lacking arms. These reflect the integration of figurative and entoptic forms during the third stage of an altered state, and provide a Chumash parallel to the many "patterned-body anthropomorphs" found in Numic territory, especially in the Cosos (see site 1). The elongation, in this case, is probably representative of bodily attenuation, one of the somatic hallucinations of a trance resulting from the sense of weightlessness and light-headedness sometimes caused by an altered state.

Other human figures are less elaborate but equally interpretable. According to the ethnographic record, headless human figures represent ghosts that are lost in the supernatural—the dead who have not successfully traveled to the land of the dead or souls that have been "stolen" by evil shamans. A small headless figure, consisting of an inverted V-shaped torso with fringe hanging down toward the cave floor, is one such figure. It is located low on the cave wall. Ghosts such as these, too, were reportedly seen by the shaman in the supernatural.

The supernatural world, then, was hardly believed benevolent or even necessarily benign. Though the shaman traveled to the supernatural to perform his various beneficial supernatural tasks, such as curing, he might also conduct sorcery in the spirit world, stealing souls and causing people to sicken and die. The potential danger of the supernatural was codified in the belief that, to become a shaman, a perilous test must be passed. Sometimes this involved battling with spirits, passing by tumbling rocks and boulders, or bypassing the fearsome guardians of the supernatural. In a similar vein, nonshamans thought there were swarms of evil insectlike creatures and sickness "inside" rock art sites, that is, in the supernatural world behind the rock walls. It is for this reason that native south-central Californian nonshamans usually avoided rock art sites, even when they were located in the middle of their village, by staying away from them, trying not to look at them whenever possible, and never touching the panels.

Perhaps not surprisingly, Redding archaeologist Eric Ritter and his father Dale recorded a tale about Painted Cave from the ranchers who had owned the land for many generations. Originally it had been told to them by an aged Chumash ranch hand. According to this indirect source, the local Chumash avoided Painted Cave during the last century, tacitly recognizing its dangerous nature. They identified certain of the motifs at the site as pertaining to dead souls or ghosts, with the centipede, in particular, thought to be a sign of death. This last reference can, of course, be interpreted at two levels. On one, the centipede, as a shaman's species, was thought capable of inflicting death. Death, on the other, was used as a metaphor for a trance due to the physiological similarities between true dying and entering an altered state of consciousness.

◼◢◣◢◣ Visiting Site 33: Painted Cave is easily found by taking the State Street exit off US Highway 101 in Santa Barbara to California Highway 154 (San Marcos Pass Road), and heading north (see map 21). You will encounter Painted Cave Road 4.8 miles north of US 101. Turn right (east) on Painted Cave Road, which will lead you up through a series of tight switchbacks around a ridge and then along the western side of Maria Ygnacio Canyon. The rock art site is located next to the road on your left, and is well marked by a California State Parks sign in a straight portion of the road shaded by oaks roughly 2.5 miles from California Highway 154. There is no parking area at the site, and no flat land anywhere nearby upon which a parking lot could be built, so you must pull over as far as possible on the very narrow rural road. There is no room to turn around here, but if you continue north on Painted Cave Road after visiting the site you will encounter East Camino Cielo Drive. Turn left on this road, which will bring you back to Highway 154. For additional information, contact the Gaviota office of the California Department of Parks and Recreation (see appendix).

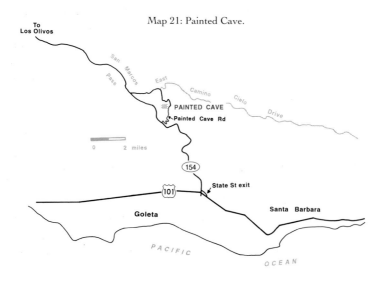

Map 21: Painted Cave.

SITE 34
Oakbrook Park Pictographs
Thousand Oaks, CA

Fishermen and Their Art

Carrizo Painted Rock (site 32) and Painted Cave (site 33) in Santa Barbara are large and elaborate examples of Chumash rock art; the Oakbrook pictograph site, in contrast, is probably closer to what the typical Chumash painted site was once like. It is located in eastern Chumash territory, in the hills above Thousand Oaks, and consists of two nearby rock shelters. These two shelters, together, comprise a few panels, each of which contains one or more simple red motifs.

Notable among these is a painting of a broadbill swordfish (photo 46), a fish that, until recently, was common in local waters. The presence of a fish in Chumash rock art at first glance might not be surprising: Living largely along the Santa Barbara Channel, the Chumash enjoyed one of the most productive marine fisheries in North America. Using canoes made from planks of driftwood, they regularly traveled from the coast to the Channel Islands, which contained large settled populations living in permanent villages. Using these same canoes, the Chumash also fished for many species of fish using handlines and nets, including both near-shore and deepwater species. Indeed, it was probably the presence and exploitation of this bountiful marine resource base that allowed the Chumash to attain a population size and density typically only found among farming peoples.

Given the importance of fish in their lives, it might be expected that fish motifs would be common in Chumash rock art. In fact, and the example from Oakbrook Park notwithstanding, precisely the opposite is the case, which serves to well illustrate the relationship between Native American foodstuffs and the species that figured prominently in shamanistic rituals and art. Throughout far-western North America there was a strong taboo against eating your spirit helper's species. This is not to say that shamanistic power could not enhance hunting powers; nonshamans often sought a spirit

175

Photo 46: Although the better-known rock art sites in the Chumash region are large and contain many pictographs, the Oakbrook Park sites contain only a handful of paintings and are probably more typical of "average" Chumash rock art sites. Shown here is a red upright swordfish, the single motif on one of the panels at Oakbrook Park. The swordfish was one of the few fish species associated with shamans. Apparently there was a special group of Swordfish Shamans who performed the swordfish dance and relied on this spirit helper. (Scale: swordfish is about 12 inches long.)

helper believed to aid their hunting abilities. But this helper was never the species that was hunted; instead it was another animal species that was associated with hunting prowess. An individual seeking deer-hunting powers, accordingly, sought the mountain lion as his spirit helper, while the owner of a Chumash canoe, hoping to ensure his success at fishing, wished for the peregrine falcon as his personal spirit helper. Furthermore, any individual who inadvertently dreamed of a common food species, such as deer, spent the remainder of his life abstaining from that food source.

Although the Chumash diet was heavily based on fish, fish and other marine foodstuffs, such as sea mammals, are relatively rare in their art, precisely because they were so important in the Chumash diet. In the few cases in which fish are depicted in Chumash rock art, however, they are usually swordfish, a species that figures prominently in their rituals and was closely associated with shamanistic practices.

The Chumash, for example, performed a Swordfish Dance as part of their annual Mourning Ceremony, which was their primary ritual, economic, and social gathering for the year. Dancers in this ceremony wore elaborate swordfish headdresses and, following south-central California ritual patterns, were almost certainly Swordfish Shamans; that is, they were shamans with swordfish as their spirit helpers. In keeping with this inference, a Chumash tale recorded early in this century recounts the belief that a group of supernatural swordfish lived inside a rock house, under the water, near Point Mugu. Another ethnographic statement tells us that, when gathering shellfish along the coast, women would throw offerings of tobacco mixed with lime—the standard shaman's hallucinogen in the Far West—to passing swordfish.

Three shamanistic themes are important to note in these two accounts. The first is the notion that the supernatural was inside a rock house; thus, when a shaman went into an altered state, and thereby entered the supernatural, the cracks in the rock at a rock art site "opened" and allowed him to enter. The swordfish near Point Mugu, then, lived within the supernatural world. The second is the underwater metaphor for trance. Because an altered state sometimes results in bodily hallucinations that are perceived as akin to being underwater, such as restricted body movements, blurred vision, and obscured hearing, shamans' visionary experiences were sometimes described verbally as "going underwater" or drowning. They were illustrated graphically by the inclusion of aquatic species (see sites 28 and 29). Although reference to this aquatic metaphor in Chumash rock art more commonly occurs through depictions of turtles, water striders, salamanders, and frogs, broadbill swordfish are the most common of the marine species so depicted. The third shamanistic theme in the two accounts concerns the tobacco offerings. Supernaturally potent species were believed to eat hallucinogens, just like shamans, who were the primary users of native tobacco. Hence, tobacco, especially tobacco mixed with slaked lime (which enhanced its potency) was offered to these fish as a tacit recognition that they were intimately conjoined with the spirit world, just like the shaman himself.

In view of all these factors, the swordfish pictograph at Oakbrook Park most likely represents a Swordfish Shaman's spirit helper, and may also allude to going underwater or drowning as a metaphor for a shaman's altered state of consciousness. Although the site has not been dated, the fragility of the underlying sandstone surface makes it unlikely that the pictograph is more than a thousand years in age, and it is probably somewhat less old. On the other hand, ethnohistorian John Johnson of Santa Barbara has radiocarbon dated a Swordfish Shaman's headdress that was excavated by archaeologists a number of years ago, having been fortuitously preserved in an archaeological site. Perhaps surprisingly, the headdress proved to be approximately 2,000 years old, suggesting that, whatever the precise age of the Oakbrook Park pictographs, Swordfish shamanism was practiced by the Chumash for at least a few thousand years.

Map 22: Oakbrook Park.

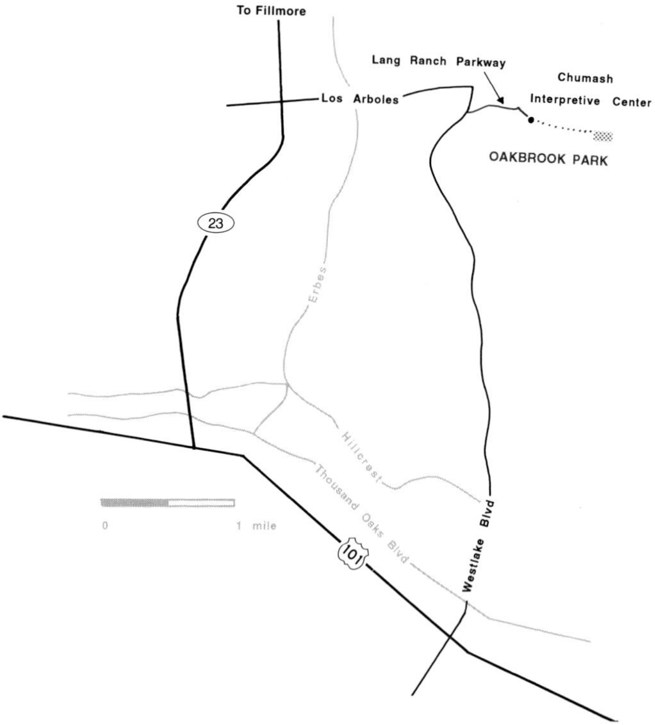

▰▰▰ **Visiting Site 34:** Viewing the pictographs at Oakbrook Park requires a walk of about 2.5 miles round-trip.

The Oakbrook Park pictograph site is located within the Chumash Interpretive Center, and may be visited as part of guided nature walks conducted on Saturdays and Sundays. To reach the center, take Westlake Boulevard in Thousand Oaks north from the Ventura Freeway (US Highway 101) for about 4 miles (see map 22). Turn right (east) on Lang Ranch Parkway and proceed about 0.3 mile to the entrance to the center, which will be on your right. For a schedule of the walks and other information, contact the Chumash Interpretive Center (see appendix for details).

MAP 108
c-4

Interior Southwestern California Region

Mockingbird Canyon Pictographs
Riverside, CA

Shamans and Initiates

As emphasized a number of times in this guide, rock art was created in at least three different contexts in southwestern California. As in the Far West generally, shamans created rock art sites at the culmination of their vision quests to record their visionary experiences, which were taken to represent events in the spirit world (see site 14). Young girls also made pictographs to conclude their puberty initiations. These, too, were intended to portray the spirit helper a girl received during her initiatory experiences (see site 17). And, in some cases, boys also painted pictographs at the end of their ritual initiations for purposes paralleling the girls', although boys' puberty art appears to have been less common than the girls' (see site 16).

Why the painting of pictographs by boys was less common than by girls is unknown: Perhaps it was simply unusual and rarely practiced, or perhaps it was more common than we realize and we simply haven't recognized many of the boys' sites. This uncertainty aside, what is clear is that the girls' sites are quite common, and are present near many large villages in southwestern California. This is particularly true in western Riverside County, where most village sites have associated pictograph panels. Less common, but still not unusual, is the presence of two or more discrete rock painting areas at or near a village. In such cases, it is common to find panels typical of the girls' puberty ceremony, as well as one or more panels of the shamans' art. Such is the case at the Mockingbird Canyon site, which contains two puberty panels (photo 47) as well as a rock shelter with a painted ceiling (photo 48), which apparently resulted from one or more shamans' ritual activities.

Photo 48: Shamanic art at Mockingbird Canyon consists of two polychrome panels containing a variety of entoptic patterns on the ceiling of a rock shelter. This portion of one panel illustrates the mental imagery of trances. (Scale: left disk motif is about 12 inches in diameter.)

Photo 47: Some sites in western Riverside County, such as the one at Mockingbird Canyon, incorporate two kinds of rock art: one is dominated by zigzag and diamond-chain motifs made at the conclusion of the girls' puberty ceremonies; the other, shamanic art, displays a wide range of motifs often painted in polychrome on cave or rock-shelter ceilings. Shown here is the approach to a vertical rock panel painted (on the backside from this image) with a few motifs from a girls' puberty ceremony. A larger girls' ceremony panel, now covered with graffiti, is located immediately across a small stream from this boulder outcrop.

The cultural affiliation of the creators of this site are not entirely clear, inasmuch as it falls near the boundary between three distinct Takic groups: the Gabrielino, who extended to the west and had, as their primary region of occupation, the Los Angeles Basin and coast; the Cahuilla, to the north and east, whose territory ran across the San Jacinto Mountains and into the Coachella Valley; and the Luiseño, whose range extended southwest into northern San Diego County and down to the coast. Nonetheless, these three groups shared numerous cultural traditions, one of which was the creation of rock art. Though the age of this site has not been determined chronometrically, Takic-speaking groups are believed to have migrated into southwestern California from the Mojave Desert only about 1,500 years ago, placing an upper limit on the age of the art.

Judging from the condition of the pictographs, an age of less than 1,000 years seems most likely.

The girls' puberty art is located on two large boulders on either side of Mockingbird Canyon Creek. You will approach the first panel on a trail coming from the north. The art is on the south side of an upright, monolithic boulder and consists of a small panel of two preserved motifs. As is typical of the girls' sites, these are red, with a diamond chain the most readily identifiable motif. As noted numerous times, diamond chains represent the rattlesnake, the preferred spirit helper for females (see sites 10 and 21).

The second girls' panel is located immediately across the creek, on a large flat vertical panel facing the stream. This panel is impossible to miss because it has been almost entirely covered with multicolored, spray-painted graffiti—a very sad commentary on how we care for our rock art sites. You will notice a bedrock mortar in front of the panel and, if you look very closely, fragments of the original red pigment in exposed areas under the graffiti.

More encouraging is the condition of the rock shelter at the site. This is located roughly 50 feet upstream from the graffiti-covered panel in a small cave created by a jumble of boulders. It consists of two areas of painted ceiling and a few motifs on the walls, The pictographs here are painted white, red, and combinations of the two; a few fugitive black lines suggest that some of the motifs may originally have been red, white, and black. Preserved motifs include patterns of dots, sets of parallel lines, concentric circles, zigzags, diamond chains, nested Us, and large grids. There is also a possible human figure, painted in red and white, on one of the ceiling panels.

Because of its location in the greater Los Angeles suburban area, Mockingbird Canyon is well known and has been widely reported in the press and popular literature. As is sometimes the case in such circumstances, certain misconceptions about the motifs have been broadcast and have now become part of the widely known folklore about the site. One of these pertains to the gridlike patterns on the ceiling of the shaman's cave. These, along with other similar "net" designs at other southwestern Californian sites, have been popularly interpreted as depictions of the Milky Way. While this would be an exciting identification if it were true, it is based on a confused reading of the ethnographic record, which states something very different. The Milky Way was in fact important to Native groups in this part of the state and, as a significant component of native cosmology, figured symbolically and conceptually in the boys' puberty initiation. However, it was represented in these rituals as a male humanlike spirit effigy woven out of milkweed fibers and not as a net as is claimed in popular discussions of this site. If the Milky Way is represented in southwestern California rock art, the ethnography is clear in indicating that this would be as a human figure and not as a grid pattern like that seen on the ceiling of the Mockingbird Canyon cave.

The motifs in the shaman's cave at Mockingbird Canyon, instead, provide a textbook illustration of the mental imagery of altered states identified by the neuropsychological model, discussed previously. Present in this cave are nearly all of the entoptic designs that commonly occur in the first stage of a shaman's trance. As I have emphasized a number of times, this confirms that the art depicts the shaman's visions, but doesn't tell us what meaning was ascribed to any of these particular motifs. Information on the meaning of these geometric patterns can only be obtained from the ethnography, and this guides us in only two areas: zigzags and diamond chains, which were universally used to indicate the rattlesnake; and spirals and concentric circles, which were whirlwinds and concentrators of supernatural power. Both are present in the cave, with rattlesnake motifs particularly common, emphasizing once again the importance of this spirit being in Native belief.

Mockingbird Canyon, then, illustrates a pattern that is relatively common in southwestern California: puberty sites on open boulders and shamans' sites in less accessible and less visible caves and rock shelters. In either case, the sites might be immediately within villages, or close nearby, yet a fundamental distinction pertained: The puberty sites resulted from public rituals and thus are easily visible, whereas the shaman's ritual activities were private and avoided by the general village populace, even when occurring in the midst of the villages themselves. In certain ways, this reflects different communicative strategies for emphasizing the same point. The sacredness of the girls' site was underscored because it was seen every day, much like a cross that Christians might place in their house. The sacredness of the shaman's site, in contrast, was even more greatly emphasized because, while universally known to be present in the midst of the village, it was avoided precisely because of its power and importance.

⋀⋀⋀ Visiting Site 35: Mockingbird Canyon County Park is located immediately outside of the Riverside city limits, in an equestrian trails park that runs along Mockingbird Canyon Creek (map 23). To reach the site take the Van Buren Avenue exit southeast off the 91 Freeway. You will encounter the Mockingbird Canyon Road stoplight on your right about 2.5 miles from the freeway. Turn right and head south down the road for about 2.8 miles, at which point you will encounter Harley John Road on your left. Take this street for a short distance. You will see a chain link fence on the opposite side of the street. Look for a break in the fence, consisting of a narrow horse gate, which is about 0.2 miles from the intersection with Mockingbird Canyon Road. Note that there are no signs, signals, or other portents advertising that this is the entry into this county park, so you must be carefully on the lookout. When you enter through this gate, you will notice a

Map 23: Mockingbird Canyon County Park.

dirt path that angles off toward your right and heads up a low hill. Take this path up over the hill and down toward the stream. About 200 yards from the gate you will near the creek and encounter a pile of boulders, which includes the upright boulder containing the first painted panel. Remember that the pictographs are on the backside of the second upright boulder from the path. The boulder with this panel is located immediately downhill from a house at 18167 Harley John Road, in case you have difficulty finding the panel. To find the other panels, cross the creek at this first panel. The second girls' panel, now destroyed, is immediately adjacent to the creek at the crossing and is covered with graffiti. Walk around this panel to the right and continue upstream for about 50 feet to the pile of boulders containing the shaman's shelter. It may take a little searching to find this shelter, but it will be worth the effort and, judging from the vandal-destroyed panel, the fact that it is a little difficult to find may be its saving grace.

SITE 36
Idyllwild Pictographs
Idyllwild, CA

MAP 109
D-9

The Cahuilla Girls' Ceremony

Another girls' puberty site is located in Idyllwild County Park, just outside the mountain town of Idyllwild in the San Jacintos. Consisting of red painted motifs, like many girls' sites, the site contains about a half-dozen zigzags, diamond chains, and variations thereof. Notable here are two finely rendered diamond chains, carefully drawn with double lines (photo 49). As with the other girls' puberty sites in this guide (see sites 17, 35, and 37), these motifs were universally identified as schematic representations of the supernatural rattlesnake, the spirit helper considered most appropriate for women (see sites 10 and 21).

The Idyllwild pictographs are located in Cahuilla territory. The Cahuilla, a Takic-speaking group whose territory ranged from western Riverside County through the San Jacinto Mountains and down

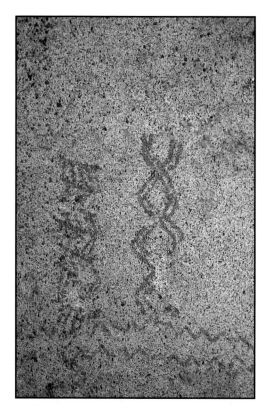

Photo 49: The relative commonness of art made during the girls' puberty ceremony is evident from the large number of such sites found in western Riverside and northern San Diego counties. A well-preserved and very accessible rock art panel resulting from this ceremony is located within Idyllwild County Park. As is typical of these sites, the dominant motifs here are zigzags, diamond chains, and variants thereof. Both designs represent the rattlesnake, the spirit believed to guard the vagina and therefore considered most appropriate as a girls' supernatural helper. (Scale: motif on the right is about 8 inches long.)

into the Coachella Valley, were responsible for a number of rock art sites discussed in this guide (see sites 14, 15, and 18). Their rock art was made by shamans, by female puberty initiates, and, probably, by male puberty initiates. It included petroglyphs, generally characteristic of the desert region in the east, and pictographs, as found at this and other sites on the western side of their territory.

Based on historical linguistic studies, Takic-speaking groups such as the Cahuilla are believed to have migrated into the southwestern California region roughly 1,500 years ago, placing a maximum possible age on their rock art sites. The region was occupied by an unknown group or groups prior to this time, who may also have made rock art. Thus, we cannot infer that all southwestern Californian rock art sites are 1,500 years or less in age, but instead must restrict this inference to those typical of the known historical patterns of rock art production such as seen at this site (see also site 38). The Idyllwild pictographs, then, cannot be older than a millennium and a half in age, but, judging from their condition and context, it is likely they are actually much younger, in the neighborhood of only a few hundred years old at the outside.

The Cahuilla girls' puberty ceremony was, in most respects, similar to the Kumeyaay (see site 17), though perhaps slightly more formalized. It involved isolation in a warmed pit for three days, thereby mimicking the ritual isolation and immobility practiced at childbirth; the ingestion of tobacco and resulting receipt of a supernatural vision; and, apparently at the culmination of initiation, the painting of the designs representing the spirit received during the girl's altered state. Following the conclusion of the initiation, the young girls were eligible for marriage and ready to assume adult responsibilities. The pictographs painted on the boulders near a girl's village, then, symbolized at once her spirit helper and also her status as an adult member of her clan.

▲▲▲ Visiting Site 36: The Idyllwild pictographs are located in a Riverside County park campground on the outskirts of the town of Idyllwild (map 24). From California Highway 243 in downtown Idyllwild, turn west on Riverside County Playground Road, which

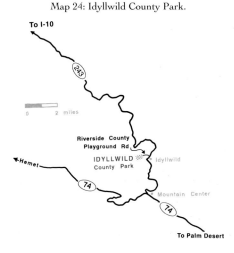

Map 24: Idyllwild County Park.

leads directly into the park. The site is on a large fenced boulder at the far end of the campground; a map at the park entrance will direct you to "Pictograph Rock." There is a small fee for day use and camping in the park. For information and camping reservations, contact Riverside County Parks (see appendix).

MAP 108
D-6

SITE 37
Lake Perris Pictograph
Perris, CA

Variations on a Theme

The relative commonness of girls' puberty sites is illustrated by the fact that three such sites are open for public visits in western Riverside County alone (see the previous two sites, 35 and 36). The third of these is located within the Lake Perris State Recreation Area in the Moreno Valley east of Riverside. It is a small site consisting of a single panel of red motifs. As is typical of the girls' puberty sites throughout southwestern California, snake symbolism is the dominant characteristic of the panel. The motifs consist of a vertical line with pendant diamond chains placed on an isolated but open monolithic boulder (photo 50).

This portion of western Riverside County, as noted previously (see site 35), is somewhat ambiguous with respect to its former Native American inhabitants: Because it is located near the boundaries of the Luiseño, Gabrielino, and Cahuilla, we cannot be certain which group historically occupied the Lake Perris area, and it is of course possible that the boundaries may have shifted over time. Cultural patterns for these three groups, nonetheless, were very similar; and we can be fairly certain that, whichever particular culture was responsible for creating this site, its origin followed the general pattern for the girls' ceremony throughout southwestern California. This involved a period of isolation in a stone-warmed earthen bed, the ingestion of hallucinogens (usually native tobacco), and the culmination of the ritual with a race to a rock, where the

Photo 50: This boulder on the southeast shore of Lake Perris contains a small panel (behind the wood and glass frame) painted during a girls' puberty ceremony. Invariably, the art at these sites was painted in red, the female color. The paintings were also usually placed on large, open vertical monoliths or vertical panel faces, reflecting the fact that the painting was part of a group ceremony, often with a number of participants and perhaps spectators.

paintings were drawn. We can place a maximum possible age on this and other girls' puberty sites at approximately 1,500 years, the time at which we believe these Takic-speaking groups first moved into southwestern California. However, given the condition of the paintings at this site, it is likely that it is only a few hundred years old.

The pictograph rock at Lake Perris is on the bicycle trail around the lake, and may be visited at any time while the recreation area is open. Another small site, located near Bernasconi Pass on the southern side of the lake, may be visited on a guided tour, which is conducted on the last Sunday of the month during the fall, winter, and spring. This site consists of a single large petroglyph and some very faded remnants of red pictographs on nearby boulders. It was identified by the ethnographer John P. Harrington as a "Taakwic puki" or "Tahquitz's House," the generic name given to shamans' rock art sites in this portion of southwestern California. Tahquitz was believed to be a very powerful shaman's spirit helper who lived on top of Tahquitz Peak, now Mt. San Jacinto, and who manifested himself as ball or heat lightning. A folktale identifies the petroglyph, which consists of two circles on either side of a vertical line, as "Tahquitz's genitals," most likely alluding to the sexual symbolism of rock art sites and the widespread belief in the great sexual potency of shamans (see sites 10 and 32).

Visiting Site 37: Seeing the Lake Perris pictograph panel requires about a one-hour walk, round-trip.

From the Moreno Valley Freeway (California Highway 215), head east on the Ramona Expressway (see map 25). Turn left onto the second road leading to Lake Perris, which is the Bernasconi Pass entrance approximately 6.7 miles from the freeway. Proceed to the Bernasconi parking area and beach. Follow the paved bike trail east, to your right as you face the lake, for roughly 2 miles (about a 30-minute walk), which should bring you to the southeastern end of the lake. You will see a large, open monolithic boulder just to the right of

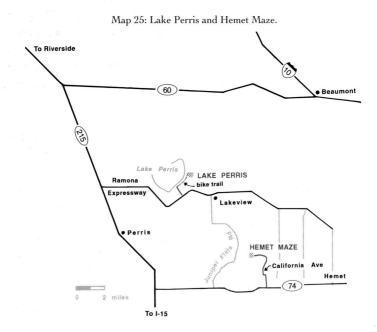

Map 25: Lake Perris and Hemet Maze.

190

the trail, with a wood frame that encloses the pictograph panel. For information about the guided tours to the other site, contact the Lake Perris State Recreation Area (see appendix). There is a day use/parking fee charged for entering the park.

SITE 38
Hemet Maze Petroglyph
Perris, CA

MAP 108
D-6

Remaining Mysteries

Throughout this guide I have emphasized what we have learned about Native American rock art by examining ethnographic studies completed early in this century, as well as by interviews with modern Native Americans. These sources have provided a very coherent and, in some cases, detailed understanding of this art, linking it to shamans, their visions, and their supernatural activities, and to the shamanistic initiations of youths. Still, it would be a mistake to assume from this that there is nothing more to learn about far-western North American rock art, and that no mysteries remain unsolved. The opposite is, in fact, the case for at least three reasons.

The first reason is chronological: Our ethnographic evidence can only be applied with confidence to rock art that is a few hundred and, perhaps at the outside, 1,000 to 1,500 years old, and we are uncertain how far back this ethnographic interpretation may be pushed. While our best hypothesis at this point is that the earlier art is also shamanistic in origin, this remains an unproven, even if imminently plausible, theory that still fails to inform a detailed understanding of the specific meaning of particular early motifs. For example, while we may identify recent Numic bighorn sheep motifs as having been produced by Rain Shamans, whether or not the same interpretation applies for bighorn engravings that

191

are 5,000 years old is simply unknown. We must always recognize, therefore, the possibility, if not the probability, that specific motif meanings may have changed over the long course of the Native American occupation of the Far West.

The second unresolved aspect of far-western North American rock art concerns the link between the art and astronomical beliefs and practices. As I noted at the outset, there has been a small but vocal group of rock art enthusiasts who find star maps, solstice observatories, eclipses, and other celestial phenomena in every panel and site. The majority of their widely broadcast archaeoastronomical claims are based on dubious logic, an ignorance of scientific method, and misreadings of the ethnographic data; and none of them have the support of any direct ethnographic evidence. That said, it is nonetheless true that a handful of the empirical observations that have been made at a few sites during the solstice, combined with the fact that Native American shamans did maintain sophisticated and detailed astronomical knowledge, nonetheless suggests a relationship between some of the art and Native cosmological and astronomical beliefs. The relationship of southern Californian rock art and astronomical beliefs and practices, then, is in my mind still unknown but worthy of serious study and may ultimately further our detailed understanding of this art. But this will only come about when and if those interested in archaeoastronomy approach this problem with scientific professionalism and rigor and a more sophisticated understanding of the Native American cultures that produced this art.

The final aspect of the unresolved issues concerning rock art reflects the standard course of scientific investigations, which almost invariably begin with a synthesis that provides the "big picture" pertaining to a particular problem or phenomenon. Once this generalizing framework has been established, researchers begin to identify, study, and understand the variability that is almost certainly present when one is dealing with human behavior. This second stage of research greatly amplifies the detailed understanding of the phenomenon in question, but it also reveals the cases in which the big picture interpretation is simply inadequate, modifying it with adjunct hypotheses and interpretations. Currently, we are just beginning to enter this second stage of rock art research in the Far

Photo 51: Although studies of the ethnographic record have greatly enlightened our understanding of far-western North American rock art, a few outstanding mysteries remain. One of these concerns the so-called maze style rock art of western Riverside and northern San Diego Counties. This example, from Hemet Maze Stone Park, is a petroglyph, but painted examples have also been found. Exactly who made these, and for what reason, remains unclear. (Scale: widest dimension of the "maze" is about 2 feet.)

West, and, as we proceed beyond the grand synthesis, we are likely to discover variability in the origins and meanings of certain sites and motifs that we heretofore had not been able to recognize or understand.

The Hemet Maze petroglyph (photo 51) is a case in point. It is a single, deeply engraved intricate gridlike maze pattern, located on an isolated boulder northwest of the town of Hemet, California. Similar very complex gridlike patterns (few are actually "mazes"), both painted and pecked, have been found in northern San Diego County, especially in the Escondido region and western Riverside County. Yet as of this writing, we are uncertain as to what cultural group made these unusual motifs and why they may have been produced.

In part, this results because they are clearly distinct from the other southwestern Californian sites that, based on ethnographic information, we know were made either by shamans (see site 14) or by puberty initiates (see sites 16, 17, 35, 36, and 37). These maze sites differ both in terms of the number and types of motifs present — often a single, very carefully rendered large "maze" on a vertical panel, versus the shamans' variety of smaller motifs on a cave ceiling or the relatively crude zigzags and diamond-chain motifs made by puberty initiates — as well as the context of the sites, which are isolated but open boulders for the mazes, versus panels in or near villages for the ethnographically interpreted art. It almost goes without saying that, lacking any idea of what culture(s) was responsible for this maze art, we do not know how old these sites may be. The maze-style sites, then, represent one of the outstanding mysteries in far-western North American rock art.

As regards chronology and cultural affiliations, two possibilities exist, each with different implications for interpretation. If chronometric dates are obtained from these sites, and if they prove to date roughly to the last 1,500 years, the implication is that these maze-style sites represent a variation in the production of the late prehistoric and historical rock art that has not yet been recognized in the ethnographic record. Instead of two potential origins for southwestern California rock art, i.e., shamans' vision quests and boys' and girls' puberty initiations, a third origin for some of the art would have to be added and, ultimately, interpreted. Alternatively, it is also possible that these maze-style sites may be older, perhaps predating the movements of Takic-speaking peoples in southwestern California at roughly 1,500 years ago, thus reflecting an ancient rock art tradition that was superseded by other practices in more recent times.

This uncertainty notwithstanding, two general interpretive inferences can still be offered. The first is that, whatever their age or culture of origin, there are strong reasons to assume this art will ultimately be tied to shamanism in some form. This seems apparent for two reasons. First, our best hypothesis at this point is that much if not all early rock art in the Far West had a shamanistic origin; and second, the motifs themselves resemble complex integrations of entoptic forms resulting from altered states of consciousness. The second inference concerns the ideological purpose of the art. While we cannot directly infer what this may have been, it probably differed from the two known ethnographic cases of rock art production, due to the differences in the context and nature of these kinds of rock art. The maze-style sites, like the Hemet maze, were formal, open public displays, differing greatly from the secretive, idiosyncratic art of the recent shamans, and even the hurried and repetitive art of the ritual initiates. The larger sociological implications of this distinction in the production of rock art are unclear, emphasizing that some aspects of far-western North American rock art are not, and may never be, unraveled.

Visiting Site 38: The Hemet Maze is located in Maze Stone County Park a few miles west of the town of Hemet in Riverside County (see map 25). To reach the site, take California Highway 74 east off the Moreno Valley Freeway (California Highway 215). You will encounter California Avenue on your left about 8.3 miles east of the freeway. Turn onto this road and follow it north to a gate that closes it off, approximately 4.0 miles beyond Highway 74. Park at the gate and walk north on California Avenue about 250 yards. You will see the Hemet Maze, which is on top of a large boulder surrounded by a chain-link fence, to your right at the end of the road. For information, contact Riverside County Parks (see appendix).

Further Reading and Other Resources

Native American Cultures of California and Nevada

Several readable and accurate books are available on the native cultures of California and the Great Basin. Although some focus on specific cultures, for general overviews I recommend the following:

Warren L. d'Azevedo, editor. *Handbook of North American Indians, Volume 11: Great Basin.* (Washington, D.C.: Smithsonian Institution, 1986). A concise and authoritative compendium describing different tribal groups of the Great Basin.

Robert F. Heizer, editor. *Handbook of North American Indians, Volume 8: California.* (Washington, D.C.: Smithsonian Institution, 1978). A concise and authoritative compendium of California tribes.

Alfred L. Kroeber. *Handbook of the Indians of California.* (New York: Dover Reprints, 1976). A classic study of native Californians. Although originally published in 1925, it remains very readable and authoritative.

Alfonso Ortiz, editor. *Handbook of North American Indians, Volume 10: Southwest.* (Washington, D.C.: Smithsonian Institution, 1983). Although primarily focused on such southwestern groups as the Pima-Papago, Navajo, and Apache, this volume also includes chapters on the Yuman-speaking groups along the Colorado River.

General Prehistory of California and the Great Basin

Two excellent overviews of California prehistory are:

Joseph L. and Kerry Kona Chartkoff. *The Archaeology of California* (Stanford: Stanford University Press, 1984). A readable summary of California prehistory for a general audience.

Michael J, Moratto. *California Archaeology* (Orlando, FL: Academic Press, 1984). An encyclopedic summary of California prehistory, including useful details on specific regions and particular sites as well as a chapter that summarizes the prehistory of southern Nevada and the Colorado River regions.

Rock Art of California and the Great Basin

A recent intellectual revolution in rock art studies has brought new dating techniques and approaches to interpreting the art that have radically changed our understanding of it. Two consequences of this revolution are, first, that most older publications are outdated and inaccurate, even though they remain widely available on the book market; and second, that most of the publications explaining these new research approaches and outlining our evolving knowledge of rock art are available only in professional archaeological literature, thus they may be not only hard to find but also very technical and difficult for the lay person to understand. The following list, restricted to books still in print and/or articles that may be found in journals at college libraries, includes some examples of both genres.

Campbell Grant. *The Rock Paintings of the Chumash* (Berkeley: University of California Press, 1965). Recently reissued, this popular book includes full-color plates of Grant's renowned copies of the polychrome Chumash pictographs and a well-written summary of Chumash culture.

————. *Rock Drawings of the Coso Range, Inyo County, California* (Ridgecrest, CA: Maturango Museum, 1968). This popular, well-written, and widely available study of the Coso petroglyphs is now known to contain incorrect interpretations and dating of the art, although if offers a good descriptive summary of these petroglyphs.

Ken Hedges. "Rock Art in Southern California," *Pacific Coast Archaeological Society Quarterly* 9, no. 4 (1973): 1–28. A good descriptive summary of the range of variability in rock art in southwestern California.

Robert F. Heizer and Martin A. Baumhoff. *Prehistoric Rock Art of Nevada and Eastern California* (Berkeley: University of California Press, 1962). Still widely available thirty-some years after its original publication, this volume offers useful descriptions of many sites in the western Great Basin; however, as with Grant's *Rock Drawings*, which borrowed heavily from the ideas presented here, the interpretations and chronologies have been discredited.

Joan Oxendine. "The Luiseño Girls' Ceremony," *Journal of California and Great Basin Anthropology* 2 (1980): 37–50. A good summary of the girls' puberty initiations in southwestern California, including their creation of rock art during the rituals.

Solveig Turpin, editor. *Shamanism and Rock Art in North America* (San Antonio, TX: Rock Art Foundation, 1994). A monograph about the origin of rock art in shamanistic beliefs and the symbolism of sites in California, Arizona, Utah, Texas, and Montana.

David S. Whitley. "Shamanism and Rock Art in Far Western North America," *Cambridge Archaeological Journal* 2, no. 1 (1992): 89–113. A summary of ethnographic evidence for the shamanistic origin of rock art in the Far West.

———. "By the Hunter, for the Gatherer: Art, Social Relations and Subsistence Change in the Prehistoric Great Basin," *World Archaeology* 25, no. 3 (1994): 356–73. A detailed study of the Coso Range petroglyphs, addressing their function and purpose in Numic society.

David S. Whitley and Lawrence L. Loendorf, editors. *New Light on Old Art: Recent Advances in Hunter-Gatherer Rock Art Research*, Monograph 36, (UCLA: Institute of Archaeology, 1994). A survey of recent advances in research, including the dating of pictographs and petroglyphs, and the use of different interpretive models to understand the art in California, the Great Basin, Wyoming, and the Great Plains, as well as the European Upper Paleolithic.

Shamanism, Neuropsychology, and Altered States

The growing popularity of shamanism as an alternative religion and philosophy, and of altered states as an approach to consciousness expansion, has increased the number of related books and magazines on the market. Those listed below emphasize studies of shamanistic practices in traditional societies and hard-science studies of altered states of consciousness.

Lowell John Bean, editor. *California Indian Shamanism* (Menlo Park, CA: Ballena Press, 1992). A compendium of professional papers on California shamanism, including Ken Hedges' good discussion of rock art as well as articles on rituals and shamanistic beliefs more generally.

Mircea Eliade. *Shamanism: Archaic Techniques of Ecstasy* (Princeton: Princeton University Press, 1951). The classic cross-cultural study of shamanism, its rituals, and its symbols. It remains widely available and very readable.

David Lewis-Williams and Thomas Dowson. "The Signs of All Times: Entoptic Phenomena in Upper Paleolithic Art," *Current Anthropology* 29, no. 2 (1988): 201–45. After explaining neuropsychological research to develop a model of the mental imagery that occurs in altered states of consciousness, this article applies the model against the cave art of the European Upper Paleolithic, arguing that it too is shamanistic in origin.

Gerald Oster. "Phosphenes," *Scientific American* 222, no. 2 (1970): 82–87. A discussion of the entoptic phenomena called "phosphenes" that occur during the first stage of an altered state of consciousness.

Ronald K. Siegal. "Hallucinations," *Scientific American* 237, no. 4 (1977): 132–40. A good introduction to the neuropsychology of altered states of consciousness.

Dating Rock Art

Rock art dating is the most technical topic considered here. Currently we may read about it only in professional monographs and journals. Those who wish to delve into such sources should see Whitley and Loendorf, *New Light on Old Art* (cited above), wherein Scott Chaffee, et al., summarize the technique used to date pictographs while Ronald Dorn discusses techniques for dating petroglyphs. Other publications pertinent to California and Nevada include:

Ronald I. Dorn. "Rock Varnish," *American Scientist* 79 (1992): 542–53. An overview for general readers of rock varnish and how it can be dated on petroglyphs, prehistoric stone tools, and landforms.

David S. Whitley and Ronald I. Dorn. "Rock Art Chronology in Eastern California," *World Archaeology* 19 (1987): 150–64. A summary of the temporal range of eastern California petroglyphs (now slightly outdated, see article below).

———. "New Perspectives on the Clovis vs. Pre-Clovis Controversy," *American Antiquity* 58 (1993): 626–47. A technical study of putative Pleistocene rock art dates and their implications for the peopling of the Americas.

Other Resources

- American Rock Art Research Association (ARARA) is composed of interested laypersons and archaeologists dedicated to the study and preservation of rock art sites. The organization meets annually during Memorial Day weekend at different locations in the West, publishes proceedings of the papers presented at annual meetings, conducts field trips, and publishes the newsletter *La Pintura*. Contact: ARARA, Arizona State Museum, University of Arizona, Tucson, AZ 85721.

- *International Newsletter On Rock Art* (*INORA*) is published three times a year in English and French by UNESCO's International Council on Monuments and Sites (ICOMOS), International Rock Art Committee. It contains summaries of recent discoveries, new books, recent rock art meetings, etc., and is the best source for information on rock art worldwide. For subscription information contact: ARARA (listed above) or *INORA*, 11 Rue du Fourcat, 0900 Foix, France.

• Piedra Pintada Books, a mail-order service dealing exclusively in books on rock art and related topics, is a great source for titles not available in chain bookstores, including some that are out-of-print. For a catalog, contact: Piedra Pintada Books, P.O. Box 1376, Claremont, CA 91711; phone 909-620-6742.

• The San Diego Museum of Man sponsors an annual rock art symposium on the first Saturday of November and publishes proceedings of the papers presented each year. Contact: San Diego Museum of Man, Balboa Park, 1350 El Prado, San Diego, CA 92101; phone 619-239-2001.

Glossary

altered state of consciousness — A trance state induced by various drugs, isolation and sensory deprivation, meditation, drumming, or other means that results in visions and hallucinations.

anthropomorph — A rock art painting, engraving, or earth figure of a human.

atlatl — A throwing board for propelling a dart or small spear, used in western North America until about A.D. 500, when bow-and-arrow technology spread.

Avikwa'ame — Spirit Mountain, or Newberry Peak, in southernmost Nevada. Center of the world and home of the creator-deity Mastamho for Yuman speakers.

band — A form of social organization consisting of a group of families loosely organized under the leadership of a headman. Membership was usually voluntary among families who generally shared the same territory and villages. The Great Basin Numic were organized in bands.

Cahuilla — A Takic-speaking native Californian group whose historical territory ranged over the San Jacinto Mountains from the Coachella Valley and into western Riverside County.

cation-ratio dating — A chronometric dating technique of the varnish that develops in the engraved area of a petroglyph. It is based on analyzing the chemical constituents of the rock varnish, which changes regularly over time.

Chemehuevi — A Southern Paiute (Numic) group who lived historically in eastern California, west and south of Las Vegas.

chiefdom — A form of social organization that usually includes classes or castes (such as rulers, commoners, and slaves) and is often led by chiefs who inherit their power. Hunter-gatherers rarely formed chiefdoms, but the Chumash organized into a simple one.

chronometric dating—Rigorous scientific techniques used to establish the precise age of archaeological samples. Radiocarbon dating is the most widely applied chronometric technique in archaeology, but a series of other methods, such as cation-ratio dating, also exist.

Chumash—A large and prosperous group who lived historically along the Santa Barbara Channel. Their territory stretched from Topanga Canyon north and west into San Luis Obispo County, incorporated the northern Channel Islands, and extended inland to the edge of the San Joaquin Valley.

Cocopa—A Yuman-speaking group who lived around the delta of the Colorado River, in northern Baja California.

conflation—A combination of different species into a fantastical composite animal. Conflations of humans and animals are common in shamanistic rock art.

Cupeño—One of the smallest Takic-speaking groups in southwestern California. Historically they occupied roughly a ten-square-mile area east of Lake Henshaw in northern San Diego County, centered on Warner Hot Springs.

cupule—Small cuplike depressions ground into rock surfaces, typically for ceremonial purposes. Smaller in size than bedrock mortars—which are large, deep conical depressions for grinding acorns, seeds, and nuts—cupules commonly occur on vertical or near-vertical rock and boulder faces. In northern and south-central California, women wanting to enhance their fertility made cupules; in northwestern California, shamans practicing weather control sometimes made them. Although the ethnography is not clear about the use of cupules in southern California and the Great Basin, their ritual use was undoubtedly related to beliefs about rocks as entrances to the supernatural world.

desert pavement—A layer of cobbles and pebbles, often darkly varnished, that forms on the ground surface in some desert areas, concealing the underlying soil.

digitate anthropomorph—A rock art motif of a human figure with prominent widely splayed fingers and/or toes.

earth figure—A motif or design, usually very large in size, made on the ground surface rather than on a cliff or cave wall or ceiling; also called a geoglyph. See also *intaglio* and *rock alignment*.

entoptic—A luminous percept generated within the human optical system. Entoptic patterns commonly occur during the initial stages of an altered state of consciousness, but they may also occur from a blow to the head, during a migraine headache, or simply by staring at a bright light for a moment and then closing your eyes. Because entoptic phenomena are generated by the human optical system, and because all humans have roughly equivalent optical systems, we often experience similar entoptic perceptions.

ethnographic record—Published documents and unpublished notes intentionally collected by anthropologists and ethnologists from elderly Native Americans, usually during the early twentieth century, in the hopes that the information would provide insight to traditional lifeways and beliefs prior to the arrival of Euro-Americans in the Far West. This record of recollections, tales, myths, customs, and so on, provides a primary resource for understanding Native American rock art. Unfortunately, rock art was never a direct concern of the information gatherers, so the details on the art are disparate, scattered, and sometimes indirect.

figurative motif—A design readily identifiable as a human, animal, spirit being, or an item of material culture, as opposed to geometric or abstract motifs; also called a *representational motif*.

Gabrielino—A Takic-speaking group that occupied the Los Angeles Basin south to Aliso Canyon in Orange County and east to about San Bernardino. The Fernandeño, centered on the San Fernando Valley, spoke a dialectical variant of Gabrielino. Some modern Gabrielino descendants refer to themselves as the Tongva.

geoglyph—See *earth figure*.

Halchidoma—A prehistoric culture, probably Yuman-speaking, that occupied the lower Colorado River Valley about midway between Needles and the Mexican border.

Holocene—A geological/paleoclimatological term referring to the Recent Period following the end of the Pleistocene or Ice Age. The beginning of the Holocene has recently been reset at 11,000 B.P. (before the present).

hunter-gatherer—A group or individual who obtains food by hunting and/or fishing and by gathering wild plants as opposed to farming.

hunting magic—A belief in using spells, charms, and/or rites to improve hunting success, either by increasing the population of the available game or by making certain animals more susceptible to the hunter. In the Far West, the only ethnographic evidence for hunting magic pertains to antelope charming among the Numic. With reference to rock art, the hunting-magic hypothesis suggested that the art was created to aid bighorn sheep hunting, an interpretation that has now been widely discredited.

intaglio — An earth figure made by scraping away the natural desert pavement, thereby creating a negative motif or design.

jimsonweed — *Datura wrightii*, an extremely hallucinogenic plant that grows in disturbed areas, such as roadsides, in southern California and Nevada.

Kawaiisu — A Numic-speaking group whose historical territory ranged from the Tehachapi region eastward to the Coso Range and Panamint Valley.

Kitanemuk — A Takic-speaking group who occupied portions of the Antelope Valley and the Tehachapi Mountains in historic times.

Kumeyaay — A Yuman-speaking group who occupied southern San Diego County, eastern Imperial County, and portions of northern Baja California. Sometimes referred to as Tipai or southern Diegueño.

Luiseño — A Takic-speaking group who occupied southern Orange and northern San Diego Counties, from the coast inland to about Lake Elsinore and Mt. Palomar.

Mastamho — The Yuman creator-deity who resided at Avikwa'ame, or Spirit Mountain, and who was visited by shamans when they entered mythic time-space during their supernatural experiences.

megafauna — Gigantic species of mammals that inhabited North America during the Pleistocene, such as the Columbian Mammoth and Smilodon, the saber-toothed cat.

Mojave — A Yuman-speaking group who lived along the lower Colorado River from below the Parker Dam area into southern Nevada.

monochrome — Painted with a single color.

neuropsychology — The integrated study of neurology and psychology.

nonshaman — Any person in a shamanistic culture who was not a practicing or acknowledged shaman; see *shaman* and *spirit helper*.

Northern Paiute — A Numic-speaking group who occupied the Owens Valley, northern Nevada, and portions of southwestern Oregon and southeastern Idaho.

nuclear accelerator radiocarbon dating — Radiocarbon dating that uses a nuclear accelerator to calculate the amount of decay of the radioactive isotope of carbon (carbon-14 or ^{14}C) within an archaeological specimen. Highly accurate readings can be obtained by measuring the amount of radioactive carbon still present in relatively small (compared to other dating techniques) specimens, allowing archaeologists to date such things such as paint pigment, which only contain a small percentage of organic matter.

Numic—A branch of the Uto-Aztecan language family that included the Northern Paiute, Shoshone, and Southern Paiute languages, and was spoken primarily in the Great Basin.

patterned-body anthropomorph—A petroglyph of a human figure with an elaborate, rectangular geometric design used to portray the body. Hallucinations of complex designs are known to occur during the third stage of an altered state of consciousness, when entoptic and figurative images combine. The patterned-body anthropomorphs apparently represent the shaman as he saw himself in the spirit world, with the geometric designs representing his patterns of power.

Paha—Southwestern Californian term common to both the shaman and the California racer snake. Rock art sites were known as "Paha's House."

petroglyph—An engraved, scratched, or pecked rock art motif.

pictograph—A painted rock art motif.

Pleistocene—The Ice Age, which ended about 11,000 years ago.

polychrome—Painted with multiple colors, such as a red human figure outlined in white.

Puebloan—Of or pertaining to the farming Pueblo and Anasazi cultures of Arizona, New Mexico, and southern Colorado.

Quechan—A Yuman-speaking group who lived along the lower Colorado River, immediately north of the Mexican border. In the popular literature, the Quechan are sometimes referred to simply as the Yuma.

radiocarbon dating—A dating technique based on the regular decay of a radioactive isotope of carbon in organic materials, such as wood, charcoal, or bone.

representational motif—See *figurative motif*.

rock alignment—An earth figure made by arranging cobbles on the ground surface into a design or motif.

rock varnish—A dark coating of tiny clay-size bits of wind-borne dust that in dry environments develops on rock surfaces such as cliff faces and within the engraved areas of petroglyphs. Sometimes called "desert varnish" or "patina."

SEM—Scanning electron microscope; a microscope that allows archaeologists to examine specimens on the micron scale.

Serrano — A Takic-speaking group who resided in the San Bernardino Mountains and foothills.

shaman — A religious functionary believed capable of entering the spirit world by going into an altered state of consciousness or trance; popularly known as a medicine man.

shamanism — A religion based on the shaman as the primary religious functionary.

Shoshone — A Numic-speaking group whose territory extended from the Coso Range/northern Death Valley region eastward across central Nevada, northern Utah, and southern Idaho, into western Wyoming.

somatic hallucination — A bodily hallucination, such as the sense of weightlessness or flying, experienced during an altered state of consciousness.

Southern Paiute — A Numic-speaking group whose territory ranged from the Mojave Desert west of Las Vegas across southern Nevada and into Utah. The Southern Paiute included the Chemehuevi in California.

spirit helper — The supernatural helper, guide, or tutelary of a shaman obtained during the course of a vision quest by entering an altered state of consciousness; often, but not invariably, a supernatural animal spirit. Nonshamans also sometimes had spirit helpers, though not as often, nor with the same degree of intimacy, as the shaman.

symbolic inversion — The use of a symbol that is logically the opposite of what would be expected in the normal course of things, which serves to emphasize the importance of the symbol through contrast rather than similarity. A good example is the female use of the rattlesnake motif, with its obvious male physical sexual allusions, paired against the male use of caves and rock shelters for rock art sites, with their physical analogies to vulvas and wombs.

Takic — A branch of the Uto-Aztecan language family spoken by a number of groups living in southwestern California, including the Luiseño, Cahuilla, Gabrielino, and others.

Tataviam — A Takic-speaking group who occupied the upper Santa Clara Valley region, between the crest of the San Gabriel Mountains and the Antelope Valley.

Tipai-Ipai — The Diegueño or Kumeyaay, a Yuman-speaking group who occupied portions of Imperial and San Diego Counties and extended into northern Baja California.

trance — An altered state of consciousness usually induced by ingesting hallucinogens such as jimsonweed and native tobacco, fasting, sensory deprivation, extreme physical exertion or pain, repetitive drumming and/or dancing, or simply by meditating. A trance, however induced, produces broadly similar neuropsychological effects.

tribelet — A form of social organization common to southern California recognizing a chief as the leader of two or more villages and having a known and named territory.

Tubatulabal — A group who resided in the Lake Isabella area of the southern Sierra Nevada. The Tubatulabal were Uto-Aztecan speakers, and thus were linguistically related, albeit distantly, to the Numic and Takic. They were the only members of the Tubatulabalic branch of this language family, suggesting that they broke off and minimized contact with other Numic speakers relatively early on, perhaps on the order of 2,500–3,000 years ago.

tuff — A relatively soft rock derived from volcanic ash.

varnish — See *rock varnish*.

varnish dating — The chronometric dating of the rock varnish that forms within the recesses of petroglyphs, as well as on stone artifacts that have been left on stable ground surfaces, and on stable landforms themselves, such as basalt cliff faces. Varnish-dating techniques include cation-ratio dating, nuclear accelerator radiocarbon dating, and the examination of the microstratigraphy of the rock varnish using a scanning electron microscope (SEM).

Vanyume — A poorly known group closely related to the Serrano, thus being Takic-speaking. The Vanyume lived along the Mojave River in the southern Mojave Desert, in the vicinity of Barstow and Victorville.

vision quest — As used in this guide, the vision quest is a systematic effort to obtain a supernatural spirit helper by entering an altered state of consciousness. It might involve a lengthy period of training, fasting, and preparation, but it might not require a retreat to an isolated or remote location.

Wanawut — A Luiseño supernatural spirit, thought to be anthropomorphic in form, who was the personified Milky Way and who was represented in southwestern California rituals by a woven human effigy.

Water Baby — A particularly potent Numic shaman's spirit helper who resides in springs and lakes and who appears in the form of a small human male with long hair. His footprints were sometimes seen at springs, and he was also sometimes called "Rock Baby" or "Mountain Dwarf."

Western Mono—Numic speakers, linguistically related to the Northern Paiute, who had moved west out of the Great Basin into the southern Sierra Nevada and who had largely adopted a native Californian rather than Great Basin lifeway.

Wukchumni—A Yokuts-speaking group who lived along the banks of the Kaweah River in Tulare County.

Yokuts—A branch of the Penutian language family, with a number of different individual languages, who lived historically on the San Joaquin Valley floor and in the foothills of the southern Sierra Nevada.

Yuman—A language family comprised of a number of different languages, centered on the Colorado River and in Baja California. It included the Mojave and Kumeyaay.

zoomorph—an animal form or motif.

Appendix

Phone Numbers

Alan Bible Visitor Center
Lake Mead National Recreation Area
702-293-8906

Agua Caliente Band of
Cahuilla Indians
Palm Springs
619-325-5673

Archaeological Conservancy
1217 23rd Street
Sacramento, CA 95817-4917
916-448-1892

BLM Resource Area Offices

    Bakersfield
    805-851-1491

    Barstow
    619-255-8700

    Bishop
    785 North Main Street, Suite E
    Bishop, California 93514
    619-872-4481

    Las Vegas
    702-363-1921

    Ridgecrest
    619-384-5400

    Palm Springs
    619-251-0812

California Department of
Parks and Recreation
Gaviota
805-682-4711

California State Parks
Lancaster Office
805-942-0662

Chumash Interpretive Center
3290 Lang Ranch Parkway
Thousand Oaks, CA 91362-4900
805-492-8076

Goodwin Educational Center
Carrizo Painted Rocks
805-475-2131

Joshua Tree National Park
619-367-7511

Lake Perris State Recreation Area
909-657-0676

Maturango Museum
100 E. Las Flores Avenue
Ridgecrest, California 93555
619-375-6900

Red Rock Canyon National
    Recreation Area (see BLM
    Resource Area Offices, Las Vegas)

Riverside County Parks
909-275-4310

Sequoia and Kings Canyon
National Parks
Three Rivers, California 93271
209-565-3341

Tule River Indian Reservation
209-782-2316

Valley of Fire State Park
Visitor Center
702-397-2088

Index

About the Author

As one of the North America's foremost experts on prehistoric art and the people who created it, David S. Whitley has written and lectured widely about the subject. He is at the forefront of groundbreaking developments in dating rock art, and he is among the relatively few professional archaeologists who incorporate ethnographic data in their research. He represents the United States at the United Nations Educational, Scientific, and Cultural Organization's (UNESCO's) International Council on Monuments and Sites.

From 1983 to 1987 Whitley served as chief archaeologist for UCLA's Institute of Archaeology, where he continues to teach. In 1994 the Institute published his professional monograph *New Light on Old Art: Recent Advances in Hunter-Gatherer Rock Art Research*. Whitley holds a Ph.D. in archaeology and an M.A. in geography, both from UCLA. He is president of W & S Consultants, a professional archaeological consulting firm in Fillmore, California.